camaro

CAMARO FIFTY

FIFTY YEARS OF CHEVY PERFORMANCE

MIKE MUELLER

FOREWORD BY SCOTT SETTLEMIRE

motorbooks

Quarto is the authority on a wide range of topics.
Quarto educates, entertains and enriches the lives of our readers—enthusiasts and lovers of hands-on living.
www.quartoknows.com

First published in 2016 by Motorbooks, an imprint of Quarto Publishing Group USA Inc., 400 First Avenue North, Suite 400, Minneapolis, MN 55401 USA. Telephone: (612) 344-8100 Fax: (612) 344-8692

quartoknows.com
Visit our blogs at quartoknows.com

10 9 8 7 6 5 4 3 2 1

ISBN: 978-0-7603-5034-8

Library of Congress Cataloging-in-Publication Data

Names: Mueller, Mike, 1959- author.
Title: Camaro : fifty years of Chevy performance / by Mike Mueller.
Description: Minneapolis, Minnesota : Quarto Publishing Group Inc., Motorbooks, 2016.
Identifiers: LCCN 2016014516 | ISBN 9780760350348 (hardcover)
Subjects: LCSH: Camaro automobile--History.
Classification: LCC TL215.C33 M84 2016 | DDC 629.222/2--dc23
LC record available at https://lccn.loc.gov/2016014516

Acquiring Editor: Zack Miller
Project Manager: Jordan Wiklund
Creative Director: Laura Drew
Cover Designer: Jay Smith–Juicebox Designs
Book Design and Layout: Brad Norr Design

On the front cover: Camaro silhouettes throughout the years. *Gabor Vida*
On the back cover: 2017 50th Anniversary Limited Edition Camaro. *GM*
On the frontis: 1969 Chevrolet Camaro Sport coupe. *GM*
On the front endpaper: At the time, the Camaro's engineering represented state-of-the-art technology. *GM*
On the rear endpaper: Chevrolet brought back the 1LE performance package for the 2017 Camaro, but this time the package was also available for V-6 LT coupes along with the V-8 models. *GM*

Printed in China

CONTENTS

ACKNOWLEDGMENTS

Once again I find so many people to thank, so few pages. Guess I really should start with my ace copy editor and the best girlfriend in the history of girlfriends, Erin Welker, who so graciously took me into her home here in Arlington, Texas, a few years back. A fellow Fighting Illini and computer geek by trade, Erin truly does know her way around the English language, as well as the kitchen. Talk about the best of both worlds. But if you do find a typo, misspelling, or grammatical blunder, blame her, not me.

The same goes for another proofreader (and my second-best girl), Ann Wolz, who volunteered to read a few pages while she recovered from injuries suffered at Arlington's AT&T Stadium, home to the Dallas Cowboys. No, she doesn't play for Dem Boyz, she guides tours at the "Jerry Dome," where the dangers apparently are just as real off the field as on. While no flags flew when she went down with a bum foot, fines and/or suspensions remain likely.

Another longtime friend of Erin's, Tawnya Braeutigam, deserves mention, too, for the groovy modeling work you can dig on page 39. Tawnya's lesser half, David, also gets a kudo or two for personally delivering his award-winning chili right to my doorstep while I slaved over this hot keyboard. Apparently he was well aware of the fact that starvation remains the biggest killer of freelance auto journalists in this country today. Hey, Dave, how 'bout pork chops next time?

Erin Welker (left), the author's significant other, took an especial interest in both Camaros and copy editing during the production of the following pages. At right is one of her better friends, Ann Wolz, who also lent literary assistance.

Dying of thirst, on the other hand, is never a risk, not with my unflappable editor at Motorbooks, Jordan "Curly" Wiklund, constantly reminding me to stop and smell the Shiner Bock. This one's for you, buddy, don't mind if I do. And while I'm at it, I'll also toast comrades up in Minnesota, Darwin "Not So Curly" Holmstrom and Zack "Hair Club For Men" Miller, who both owe much of their increasing sunscreen usage (on their ever-growing scalps) to my interpretation of "deadline." C'mon, if the idea wasn't to push you guys toward your graves, why does it have "dead" in it? I blame Webster.

But seriously, folks, where might I find myself if not also for so many gracious General Motors people in and around Detroit? Peggy Kelly and Larry Kinsel continue to prove that no hoop is too high, no archival photo request too obscure—save, perhaps, for those shots from DeLorean's prom. Maybe next time, huh? Similar accolades go to GM Heritage Center's able-bodied archivist Christo Datini, a researcher whom you can't stop, you can only hope to contain. If only the same could be said about Verlander, Cabrera, and the Tigers, right Christo?

I can't offer up enough thanks to Monte Doran (Chevrolet Communications), Tara Kuhnen (Chevrolet Account Supervisor), and Pamela Flores (GM Design Communications) for repeatedly dropping everything to address my pleas. Pam especially went beyond the call of duty, twice even responding to an e-mail over the weekend. What's the world coming to?

Just ask another old friend in Motown, Tom Read. When this project started, his shingle still read "GM Powertrain Communications." Then, bam, I found myself knocking on a door marked "GM Global Propulsions Systems Communications." Not only does Tom now carry a two-page business card, he also finds he can't get away from me that easy. And where can he go from here? Just askin'.

I owe Monte Doran a steak and/or Lowenbrau for putting a 2016 Camaro at my disposal for a last-second photo opp—which never would've happened if not for Mr. Can-Do, Jesus Pagan, and his hard-working press fleet team at Event Solutions, in Carrollton, Texas. Trust me, Jesus, I'll learn to read a calendar next time.

Former GM personnel also stepped up to share a story or two, including engineer Denny Davis and Harry Turner, a segment manager back when fourth-generation Camaro development was underway. Then there's the fabled "Fbodfather," Scott Settlemire, the Camaro man to end all Camaro men, whose knowledge of the breed is only exceeded by his willingness to share all he knows—and with unbridled enthusiasm, to boot. You can learn more about this

GM Global Propulsions Systems Communications man Tom Read (at right putting the torque to Chevrolet's 100 millionth small-block V-8 in November 2011) made the mistake of picking up the phone a dozen years ago to find this book's author on the other end. The Mueller hotline hasn't stopped ringing since. ***GM***

hopelessly devoted pony lover by simply turning this page, as well as a couple hundred more.

Various veteran journalists also stepped up unswervingly when I came calling, including *Car and Driver's* Don Sherman, fellow Motorbooks author (and former *Autoweek* staffer) Larry Edsall, and longtime drag racing photographer/author Bob McClurg. Former editors Jim McCraw (*Super Stock & Drag Illustrated*), Steve Statham (*Musclecar Enthusiast*), and Tom Shaw (*Muscle Car Review*) once again were there when I needed them, as was *Autoweek* creative director Ken Ross and Crain Communications (*Autoweek* and *Automotive News*) archivist Corinne Young.

I additionally can't forget all the learned guys at the Camaro Research Group (www.camaros.org), nor overlook their contributions to much of the first-generation history recounted here. See page 30 for a list of additional websites that, like those CRG pages, simply have to be visited if you call yourself a Camaro guy or gal. And while we're talking essentials, I have to pay homage to authors Gary Witzenburg and Michael Lamm, whose classic books (*Camaro! Challenger to Champion* and *The Great Camaro*, respectively) represent must-haves for any research library, let alone one dedicated to Chevy's pony car.

Locally, my thanks go to Shawn Moran, Ken Jacobs, David Goetze, and all the folks at the North Texas Camaro Club (www.northtxcamaros.com) for wasting little time coming to my aid. Crossing state lines, I simply have to mention my great uncle (no, really, my mom's brother is truly great), Sam Hart, in Mt. Vernon, Illinois, and his son, retired USAF Lt. Col. Rodney Hart, in Colorado Springs, Colorado, for their assistance getting my U-2 chase car story (see page 194) off the ground. Mucho appreciation also goes up Wisconsin way to Mecum Auctions (www.mecum.com), home to one of the best car guys I know, consignment director John Kraman.

Last—but no way least—I extend laurels and hearty handshakes nationwide to the various Camaro owners who allowed me to photograph their rides dating back to my earliest days in this business, circa 1983. Though I can't mention all of you within these tight confines, I'll always remember your contributions as well.

And I'll gladly tip multiple Shiners to ya'll, too.

FOREWORD

"The name CAMARO means 'friend,' 'pal,' or 'comrade.' Thus, it suggests the real mission of our new automobile—to be a close companion to its owner, tailored to reflect his or her individual tastes and at the same time provide exciting personal transportation. Chevrolet has chosen a name which is lithe, and graceful—and in keeping with our other names beginning with 'C.' Thus, it suggests the comradeship of good friends . . . as a personal car should be to its owner."

These words were spoken by then Chevrolet general manager Elliot M. "Pete" Estes on June 29, 1966, when the name Camaro was revealed to the press all across America. Over the last fifty years, we at Chevrolet have worked hard to remain true to the brand promise and to Mr. Estes's vision—with style that turns heads, handling like it's on rails, and offering affordability to the average guy or gal.

Camaro has become "a part of the family" to millions of owners. It's been the subject of songs, and it has starred in many movies and TV series. Nearly everyone has a Camaro in their past. If they didn't own one, then their parents, siblings, or best friends had (or have!) one, and they reflect fondly about the good times shared in a Camaro. We're honored to know that many people name their Camaros, and that they make owning a Camaro a lifestyle. From Camaro Clubs to Camaro enthusiast websites, there is always something and someone to reach out to with questions or comments about the Camaro.

I often wonder if Mr. Estes had any idea that day back in 1966 that a car by the name of Camaro would bring like-minded people together all over the world in true Camaro "comradeship" . . .

And, finally, Camaro has been a passion of mine since September of 1966 when I first laid eyes on one. Growing up in a family-owned dealership allowed me to watch the Camaro progress through the generations until I was selected to be a part of Team Camaro in 1996. It's been a rare and true privilege to be one of the many custodians of the brand and to ensure that we bring you the best possible Camaro.

Now sit back, relax, and read all about this wonderful automobile. Thanks for your passion for all things Camaro!

Scott Settlemire
Grosse Pointe Farms, MI
April 2016

Noted Camaro devotee Scott Settlemire (center) was on hand to introduce the Camaro convertible concept to then California governor Arnold Shwarzenegger (left) at 2007's Los Angeles Auto Show. To the right is auto show director Andy Fuzesi. *GM*

TEXAS
67 Z28

Introduction

A CHEVY FOR THE AGES

Chevrolet's beloved Camaro surely qualifies as an American automotive icon and remains to this day one of the most exciting high-performance cars available at any price. On rare occasions, a car comes along that shapes its time; it defines the era in which it existed. Chevrolet's Camaro was just such a car. It was what marketing types today call a "category killer," a car that is so good and so right for the market that it sucks out all the oxygen, leaving competitors gasping for breath.

Cars can become category killers for a number of reasons. Usually it's because they utilize technological advances that make competitors' cars obsolete. The first example of this was Ford's Model T, and the technology was mass production, which enabled Ford to sell its Model Ts at prices so low that they undercut even motorcycle prices. This made Ford's car accessible to masses of people who could never dream of affording any other vehicle. Ford's flathead V-8 engine also brought affordable performance to the masses so successfully that no other manufacturer even fielded a competitive flathead V-8.

But side-valve technology was a dead end even in 1932 when Ford introduced its V-8. The introduction of tetraethyl lead as a fuel additive in 1923, which made possible the higher compression ratios that favored an overhead valve design, meant the side-valve design's days were numbered. This set the stage for Chevrolet to introduce an engine design so good that it remains a category killer to this day: the overhead-valve small-block V-8. When Chevrolet dropped its advanced small-block on the motoring public for the 1955 model

(opposite)
Six generations and a half century later, the Camaro is still running strong as ever. Leading the way in 1967 was the original Z28 (front), a small-block screamer intended to take the Trans-Am road-race circuit by storm. To the upper right is the first of the latest breed, a 2016 LT coupe, a V-6 model more than capable of changing the way customers looked at economical transport.

year, it produced one of the most iconic engines of the twentieth century, one that is still going strong in today's phenomenal sixth-generation Camaros.

And when Chevrolet introduced its stunningly beautiful 1967 Camaro, it produced another category killer for the performance-coupe market. Competitors would certainly field worthy cars for the category, but the Camaro would stand the test of time. Most others have come and gone, and few ever came close to matching the Camaro's success in the marketplace. The Camaro had its ups and downs—as do all iconic nameplates—and even took a brief hiatus in the early years of the twenty-first century. But you can't keep a car that good down.

Made up of sport coupes only, Camaro's second generation began midyear in 1970 and carried on through 1981. Pictured here is 1970's Z28, powered by a Corvette-sourced 360-horsepower small-block. *Tom Shaw*

In August 2006, GM chairman and CEO Rick Wagoner announced that Camaro would return; this after visitors to Detroit's North American International Auto Show had gone ga-ga over a sensational concept vehicle the previous January. "The overwhelmingly enthusiastic response to the Camaro concept continues to remind me of the uniquely iconic place our products have in customers' hearts," said Wagoner. "Camaro is much more than a car; it symbolizes America's spirit and its love affair with the automobile."

Chevrolet began taking Camaro orders for 2010 on October 13, 2008, kicking off the model's fifth generation. "[It] will be almost identical to the

FROM THE HORSE'S MOUTH I

"While Camaro has evolved with cultural and design trends, it has consistently represented the approachable, attainable sports car with an emphasis on fun that's evident in its styling. As a result, each generation of Camaro has evoked an emotional connection with enthusiasts—connections spanning forty-eight years and five generations."

GM Global Design Vice President
Ed Welburn, February 2015

Ed Welburn became GM's sixth design vice president in October 2003. In March 2005, he was named the corporation's first vice president of Global Design. *GM*

A MATTER OF STYLE

When Chevrolet unveiled its reborn LT1 V-8 for 1992's Corvette, officials were quick to point out the missing hyphen—the decal last seen on forward-hinged fiberglass hoods twenty years before clearly read "LT-1." But both usages, with hyphen and without, appeared in print back in the 1970s, thanks in part to Chevy's own mixed messages. Official RPO listings used "LT1" for the engine, while the cars themselves were promoted as "LT-1" Corvettes. Before that, RPO Z28 also morphed into "Z-28" when the 302-powered Camaro was introduced in 1967, then was recast yet again as "Z/28" after badges went onto this hot pony the following year. Later, the advertised name went full circle, back simply to "Z28," the reference used throughout this book (until Chapter 9) in the best interest of continuity. Similar varying usages commonly appear whenever historians write about Chevys, hence the prevailing style seen here: no hyphens, just as official options paperwork always spelled it out.

The same rule-of-thumb applies to engine/transmission nomenclature. You'll see both "Turbo-Jet" (on air cleaner decals, most prominently) and "Turbo Jet" mentioned in official paperwork, and Turbo Hydra-matic citations over the years have commonly included "Turbo-400" and "Turbo 400." Or "TH-400," "TH 400," and even "TH400." Again, titles sans hyphens will be the style seen in this book, if only for simplicity's sake. Historical consistency also comes into play concerning Chevrolet's long-running "mouse motor," now touted formally from corporate sources as the honored "Small Block." But, while it does deserve such capital treatment, you'll see "small-block" on these pages, per long-standing tradition. Funny thing, though; when the company's world-famous logo is mentioned in GM press releases, a lower-case "b" is almost always present. Capitalization, however, is the order of the day here. After all, it isn't just anyone's neck wear, it's Chevy's Bowtie.

concept, a thoroughly modern interpretation of the 1969 model, considered by many to be the best design of the car's first generation," promised Ed Welburn, GM's global vice president of design—and, coincidentally, the proud owner of a '69 Camaro SS.

A second-generation Camaro ran from 1970 to 1981, a third gen from 1982 to 1992, and a fourth from 1993 to 2002. The first class debuted before the automotive press in September 1966 to compete with Ford's Mustang, unleashed two Aprils before.

Henry Haga, in Chevrolet Number Two Studio already had various sketches done in August 1964, when general manager Semon E. "Bunkie" Knudsen was given the order to develop a new sporty coupe.

Curiously, a few years before, Knudsen had been instructed by the same bosses to decline chief

continued on page 16

Introduced for 1982, third-gen Camaros initially were all hatchback sport coupes but could be transformed into wind-in-the-hair travelers by ordering RPO CC1, the T-top option. A truly topless body returned for 1987 via conversions made by ASC, Inc. A CC1-equipped Z28 from 1984 shows off its 305-cid V-8 here. *GM*

THE BIG PICTURE

While Camaro's parent company celebrated its 100th birthday in 2011, the Bowtie legacy actually dates back to September 1908, when former carriage maker William Crapo Durant established the General Motors Company, which at first encompassed the fledgling firm named after engineering pioneer David Dunbar Buick. The Buick Manufacturing Company, founded in 1902, had initially floundered before Durant came aboard to right the ship in 1904. Durant then used Buick to get GM rolling, and quickly brought Oldsmobile into the fold late in 1908. Durant also acquired Cadillac and Oakland in 1909, with the latter firm later evolving into Pontiac.

Although concerned bankers (alarmed by his rampant expansionism) booted him out on his wallet in 1910, Durant bounced right back, opening Chevrolet in November 1911. His goal was to beat Henry Ford at his own mass-production game, an approach not everyone involved agreed with. From the moment Durant hired him, famed racer Louis Chevrolet envisioned his moniker emblazoned on the noses of higher-priced, high-performance automobiles. But the boss only had eyes for Henry's low-priced, history-making Model T, which had dominated the market since its introduction in October 1908. Chevrolet finally quit in disgust in September 1913, leaving his surname behind, bound rightfully by contract in Durant's tight clutches.

Born December 25, 1878, Louis Chevrolet began his celebrated racing career in 1905, then hooked up with Billy Durant six years later to form the Bowtie brand that still stands tall a century-plus later. He died in Detroit on June 6, 1941. *GM*

Floundering when William Durant came aboard in 1904, Buick was transformed into America's leading automaker within four years. Using this revived firm as a foundation, Durant then opened General Motors in September 1908. Durant rests here (in the white cap) in a 1905 Buick Model D competing in the American Automobile Association's 1906 Glidden Tour. *GM*

Billy Durant, of course, was right; Chevrolet Motor Company was better off building affordable automobiles for the masses. Company coffers were already filling up nicely when the Model 490 appeared, in prototype form, in January 1915, taking its name from its tidy price tag, $490, same as Ford's Model T. Affordability guaranteed success; within seventeen days of going on sale (as a 1916 model) on June 1, 1915, the 490 inspired $23 million in orders.

Total Chevy production for 1916 jumped up by nearly five times to more than 70,000, and annual sales the following year surpassed 125,000, a bonanza that allowed Durant to buy a return ticket back into GM offices via stockholder channels. With 54.5 percent of corporate stock in hand, he marched in and retook control in September 1915, then made Chevrolet a GM division in May 1918. Further expansion followed as Durant captured more than thirty other companies during the Teens. By 1929, GM was offering seventy-two different models, including both cars and trucks.

The rest is history, and so was Durant after an economic recession forced GM stock to nose dive in 1920. He was shown the door a second time, never to return to such lofty heights. Another much smaller Durant empire rose again but was defeated by the Great Depression during the early 1930s. General Motors, meanwhile, simply shrugged off such troubles and continued its offensive against entrenched leader Ford, as well as a new combatant, this one founded by Walter P. Chrysler in 1924.

New leadership at GM considered killing off Chevrolet after Durant's departure, but Alfred Sloan Jr. would have none of that. Sloan, the big man brought in to sort through the mess left behind by GM/Chevy's founding father, felt it was the remaining "Durant officers" who deserved the axe. Along with cleaning house, he also redirected the Bowtie division towards a more promising future, a new niche where Chevys cost a little more than their rivals from Dearborn but those extra dollars bought a little more refinement, a little more comfort, a little more car.

Making former Ford production manager William Knudsen Chevrolet's new president and general manager in 1924 really helped to get things rolling. Knudsen's immediate aim was to unseat Ford atop the Detroit sales leader board, and he wasted little time scoring that bull's-eye. Ford sold eight vehicles for every one Chevrolet in 1924, but that ratio dropped to 4:1 in 1925 and to less than 3:1 in 1926. Chevy then became numero uno for the first time in 1927, though it must be said that Henry Ford contributed partially to this achievement by shutting down his company temporarily that year in order to reorganize production lines for his highly anticipated Model A.

Knudsen's gang topped the charts again, by far, in 1928, but Ford regained the lead the next year after the Model A got warmed up. A truly tough battle ensued as Chevrolet by then had its own milestone machine—the "Cast Iron Wonder," a new overhead-valve six-cylinder engine—up and running.

Introduced just before Thanksgiving in 1928, the Chevy six was a real winner, a plain fact Henry Ford quickly discovered. Knudsen had first ordered his engineers to trade four cylinders for six after it he'd learned that Ford had his new model in the works. Tit for tat. The new six was even ready for 1928, but Knudsen chose to hold back and let the Model A have the stage first. Although Ford's A-lister starred in 1929 and '30, the six-shootin' Chevy stole the show in 1931, and again in 1932, even after ol'

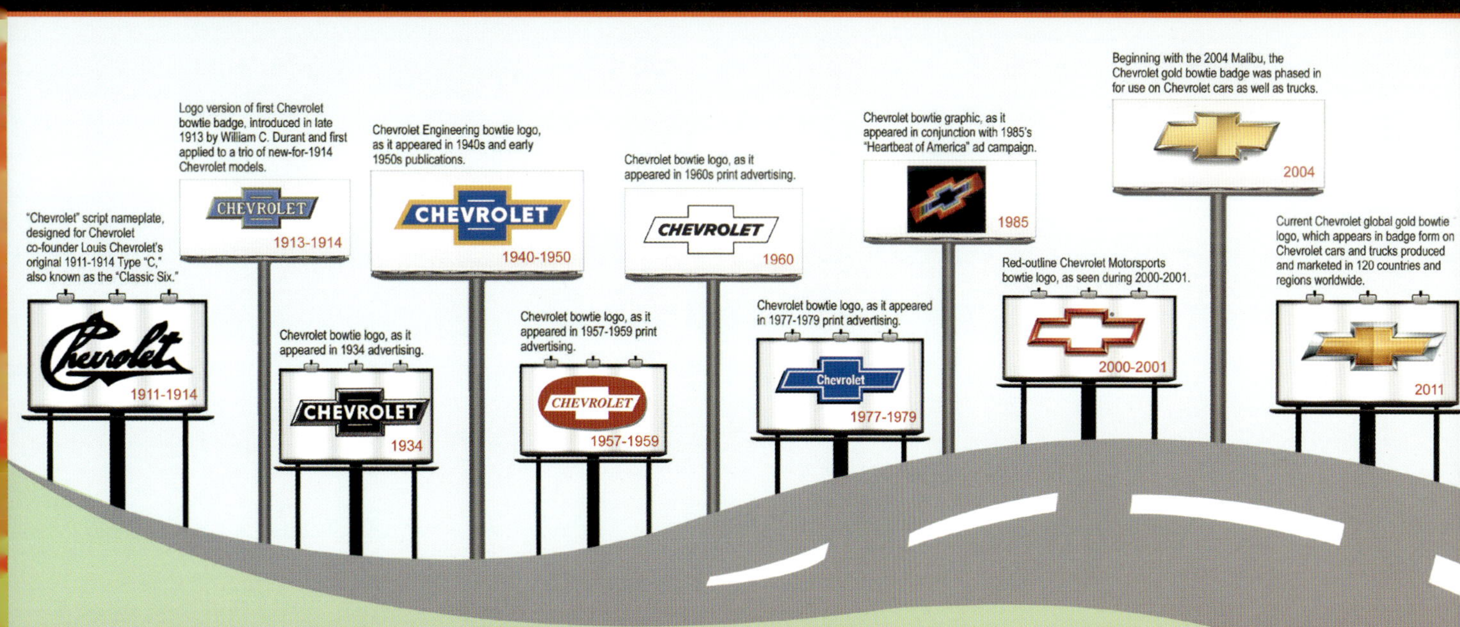

First seen in 1913, Chevrolet's world-famous Bowtie logo has undergone its fair share of updates over the years. *GM*

Henry made more history by introducing his "flathead" V-8. While Dearborn temporarily regained the lead in 1935, it was Chevrolet that totally dominated from there. The Bowtie brand was still riding high behind that renowned "U.S. 1" banner twenty years later when its own ground-breaking V-8 finally emerged to overshadow the old, reliable "Stovebolt."

So where did that world-famous "Bowtie" come from? Some claim it was morphed from the simple white cross on the Swiss flag, in honor perhaps of Louis Chevrolet, who was born in Switzerland. History, however, generally credits William Durant with the idea.

"It originated in Durant's imagination when, as a world traveler in 1908, he saw the pattern marching off into infinity as a design on wallpaper in a French hotel," explained *The Chevrolet Story of 1961*, a publication issued by GM to help mark Chevy's 50th anniversary. "He tore off a piece of the wallpaper and kept it to show friends, with the thought that it would make a good nameplate for a car."

Other stories have also been told. In her 1929 book *My Father*, Durant's daughter, Margery, spoke of dad's habit of doodling during dinner. "I think it was between the soup and the fried chicken one night that he sketched the design that is used on the Chevrolet car to this day." Later, in a 1973 interview, Durant's widow, Catherine, spoke of how her husband had noticed a similar shape in a newspaper ad while on holiday in 1912 and instantly proclaimed how great it would look on a Chevy.

It sure did. Reportedly appearing for the first time in the October 2, 1913, edition of the *Washington Post*, Chevrolet's Bowtie still stands as one of the world's most-recognized corporate logos.

During aerodynamic testing, engineer Richard Quinn noticed that the 2014 Z28's Bowtie badge was preventing cooling air from reaching the radiator. Removing it was out of the question, so he simply cut out the badge's center section, resulting in the functional "Flowtie." *GM*

continued from page 13

designer Irv Rybicki's idea for a gorgeous "personal coupe" based on the compact Chevy II, with the rationale being that the division didn't need any more models, not with Chevrolet's fifth, Chevelle, then being readied for its 1964 debut. As Rybicki later recalled, his promising proposal mimicked much of the Mustang's makeup at a time when no one at GM even knew Ford's milestone was in the works. But it was not to be.

Just five years before, the Bowtie parade featured only two floats, with the flashy Corvette leading the way, well above the mainstream, for the full-size Impala/Bel Air/Biscayne trio. Excusing Chevy's 'glass-bodied two-seater and Ford's personally luxurious T-bird, the Big Three didn't start offering multiple

A hatchback coupe shell again was the norm when the fourth-gen Camaro debuted in 1993. A factory-direct convertible then joined the lineup the following year. A 1967 SS 350 droptop shares the stage here with 1994's topless Z28.

model lines to the masses until late in 1959, when Corvair countered American Motors' Rambler (introduced the year before) in Detroit's burgeoning compact competition. Ford immediately followed suit with Falcon, as did Chrysler Corporation with Valiant. In 1962, Dearborn expanded its lineup again, this time with a repackaged Fairlane running midway between its full-size Galaxie and compact Falcon. Bingo, the birth of the intermediate mid-size car.

At first, GM didn't respond to Fairlane, but it did have its "senior compacts," Buick's Special/Skylark and Oldsmobile's F-85/Cutlass, each introduced for 1961 with a little more size, a bit more engine and a lot more prestige than Chevy's rear-engine Corvair. A small V-8, cast of aluminum, was standard up front for these two upscale small cars.

Also new for 1961 was Pontiac Motor Division's "sophomore" compact, Tempest, a true budget buggy in base form. Standard power came from a slant-four motor, created curiously by halving Pontiac's 389-cid V-8. Buick's aluminum engine was a rare Tempest option in 1961 and '62 before being superseded by PMD's own iron V-8 the following year.

More car than Corvair, but much less compared to the Buick-Olds-Pontiac (B-O-P) triumvirate, Chevy II debuted (also with a standard four-cylinder) for 1962 and served to put Knudsen in charge of four distinctly different model lines. The midsize Chevelle filled a noticeable price gap between the big-car

GM people claimed they'd already invented the sport compact segment when Ford rolled out its first Mustang in April 1964. Introduced for 1960, Chevrolet's ground-breaking Corvair was available in various forms, from the yeoman four-door sedan shown here to the carefree Monza coupe. A Monza convertible debuted midyear in 1962. *GM*

PONY PROGENITORS

Lee Iacocca and the rest at Ford knew they had a hot seller even before Mustang started kicking up dust on April 17, 1964. Initial forecasts claimed about 100,000 would be sold that first year, but Iacocca was far more optimistic. He had greater goals in mind, specifically Detroit's record for new-model first-year sales, set by Ford's own Falcon four years before. Iacocca's battle cry became "417 by 4-17"—the plan was to sell at least 417,000 Mustangs by 4-17-65. No problem. By April 17, 1965, the total was nearly 419 grand.

Of that record-breaking tally, 121,538 were so-called "1964½" cars. Introduced in the middle of a model year, the first Mustang experienced an extended eighteen-month run that typically ended in August 1965. Most sources initially tagged them all with "1965" labels. But various running changes differentiated the cars built before August 1964 from those assembled afterward, thus making it relatively easy to group the Mustang's early production run into 1964½ and true 1965 categories. Identifying the differences is not possible in twenty-five words or less, so you'll have to settle for the most readily recognizable clue: 1964½ Mustangs used generators, '65s used alternators. The production count for the "true" twelve-month 1965 model run was 559,451 Mustangs.

Few people remember that Plymouth's Barracuda actually beat Mustang out of the gate in April 1964. At left is a Formula S from 1966; at right is the last of the breed, 1974's 'Cuda.

Like Camaro, arch-rival Mustang also is presently rolling through its sixth generation. A 1968 convertible (right) represents the first generation, which lived from 1964 to 1973. At left is a fifth-gen California Special from 2008. Demonstrating the latest platform (middle) is a 50th Anniversary GT.

Funny thing, though: today, few remember that Chrysler actually beat Dearborn to the punch in the spring of 1964. Nearly two weeks before *Time* and *Newsweek* editors began stumbling all over themselves to get their scoops on the new Ford, Plymouth's Barracuda debuted to next-to-no fanfare on April 1. So why, then, didn't it inspire the market segment name we all still use without question today?

For starters, consider sheer impact. As mentioned, Ford set all kinds of sales records with its original Mustang, rolling more than a million off the line by March 1966. Plymouth let about 23,500 Barracudas loose in 1964, followed by 60,000 in 1965, and 36,000 in 1966. Not bad numbers, mind you, but a history-maker this Plymouth clearly wasn't.

Furthermore, a rocket scientist wasn't required to determine that Barracuda was a knockoff. After learning of Ford's plans, Plymouth people had hastily prepared a copy, speeding the Barracuda through short-order development in less than two years to hopefully upstage

Iacocca at his own game. Making this rapid response possible was the plain fact that Plymouth designers more or less took the division's existing A-body Valiant and melded on a sleek, sloping rear window. From the cowl forward and beltline down, the original Barracuda was essentially all garden-variety A-body, complete with a "Valiant" badge on its tail. But setting it miles apart from its plainly dull sibling was that huge rear window, then the largest expanse of glass ever installed on an American car.

Bucket seats came standard up front, but basically all other features were Valiant carryovers, including suspension (front torsion bars, real leaf springs) and powertrain parts. All 1964–66 Barracudas also were sport coupes—a convertible didn't arrive (along with a "notchback" hardtop) until the breed was redesigned for 1967.

So it was that Ford defined this new breed. Be the name Camaro, 'Cuda, Firebird, or Javelin, these long-hood/short-deck models have always been and will always be known as "pony cars," not "predator fish cars."

plateau and Chevy II's low-buck perch, and in turn helped Chevrolet blanket the market during the 1960s like no other car company before. Nor since.

But Mustang still owned the road as far as this country's rising youth market was concerned a half century back. Industry-watchers unclear on this fresh-faced demographic at the time needed only to do the math. Start with 1946, add sixteen years, and what do you have? Try an ever-growing army of eager, recently legalized drivers, offspring of the hundreds of thousands of lonely soldiers who returned home after World War II ended. The "baby boom" those veterans bred promised a stampede of young Americans during the 1960s, all ripe for the pitching. By 1970, fifty-two out of one hundred US citizens were younger than thirty, a reality not missed by forward-thinking marketing moguls.

"While the kids don't spend all the money in this country, they are the ones everybody is playing to," explained Ford Motorsports director Jacque Passino in 1970 in a *Motor Trend* guest editorial. "The miniskirt wasn't invented because the little old lady from Pasadena wanted one. But she's probably wearing one today, because it's all she can get. Some large chain department stores don't even stock clothes for the over-30 set anymore."

Lee Iacocca was Detroit's first merry marketeer to embrace boomers when he fathered Ford's Mustang. Both sporty and truly affordable at the same time, America's original pony car attracted these youngbloods (along with those young at heart) like vampires to a virgin. But bucket seats and a floor shifter alone didn't quite warm all

Pontiac introduced Americans to an undeniably new breed in the fall of 1963. The GTO not only kicked off Detroit's modern muscle car era, it established a new-model sales record at Pontiac. At left is general manager Pete Estes, who moved over to Chevrolet in 1965. *GM*

Chevrolet's fourth model, Chevy II, joined Corvair, Corvette, and the full-sizers (Impala, Bel Air, and Biscayne) in 1962. At the top of the Chevy II line was the upscale Nova, offered in sedan, station wagon, sport coupe, and convertible forms. *GM*

customers' corpuscles. Some wanted a sporty car to run like a sporty car, not just look and feel like one. Some had also already noticed just how fun, and not to mention fast, moving from point A to B could be.

Countless young boomers watched while teens of the 1950s had scared their families half to death, as naysayers saw it, by hot rodding their way to hell. But despite narrow-minded predictions from right-wing-thinking parents, priests, and politicians, the fire-breathing hot rod—along with the paganistic rock 'n' roll blaring from its radio—did not help bring about the fall of civilization as we know it. On the contrary, it inspired an entirely new industry, the aftermarket performance parts biz, which, as the 1960s receded, was raking in more than $1 billion annually.

"It's the much talked about youth market that has pushed sales to such heights," added Passino in 1970. "Statistics show that today's kids like cars. Economists estimate that teenagers and those just out of their teens spend $3.5 billion annually on automobiles and automotive items. Automobiles are no different than clothes. The kids of today want an 'in' look in the things they wear.

Along with dominating Trans-Am racing in 1968 and 1969, first-gen Camaros predictably made mucho hay in stock-class drag racing, with quarter-mile legend Bill "Grumpy" Jenkins doing the bulk of the baling. *Bob McCurg*

PONY TALE

Car Life editors are commonly credited with first calling a horse a horse, of course, this after introducing "Ford's hot new pony" in 1964. The term "America's Pony Cars" then started showing up on *Car Life* pages after Camaro made the scene three years later, and the die was cast. While Detroit's official label for this segment initially read "sport compact," on the street they were all pony cars, be the badge Dodge, Plymouth, AMC, Mercury or Pontiac. And this term of endearment remained, regardless of how many times this market niche moniker has morphed over the years. The "compact" reference lost meaning by the 1980s—this after truly small small cars began making Camaro and Mustang look humongous in comparison—yet "subcompact" still became the published norm for both rivals during this decade. "Sports car" later gained favor, especially after even base V-6 models began proving really fun to drive. But, by 1998, not all Chevy enthusiasts were on the same page. Consider what Camaro brand manager Dick Almond had to say in August that year about 1999's upcoming F-body: "After reengineering this car specifically for the sport coupe market, we have a vehicle that not only looks like a muscle car should, but performs like one, too. Camaro has proven itself as an honest sports car that's performance-oriented, yet more affordable than many exotic sports cars. In other words, it has stayed true to its 'pony car' roots." Sport coupe, muscle car, sports car, pony car—which was it, Dick? Chevy officials these days prefer sports car or performance car when classifying their latest Camaros. But we'll stick with pony car, the long-running favorite.

John R. Bond's *Car Life* magazine is commonly credited with popularizing the term "pony car."

They also want the 'in' thing with regard to cars. This has led to the demand for the so-called muscle cars and has helped create the tremendous growth pattern in the accessory business."

Clearly, the pump was primed in 1964, and Mustang wasn't the only "in thing" flooding the market that year. Like Iacocca, Pontiac Chief Pete Estes had eyes; he could see that the boomer set might also eat up an affordable, mass-produced factory hot rod, with or without their spoons. Enter GTO, the midsize milestone that got Detroit's original muscle car era rolling.

By 1966, everything was in place—a young-leaning, thrill-seeking target market; hip music; good vibrations—and everything was cool, save for the economy, which had heated up considerably. Perhaps most importantly, cars were truly groovy, and the hotter the performance, the groovier the car. It just might have been the best time ever to be young. It certainly was a great time to be building automobiles—and/or one-upping the competition.

continued on page 24

MILESTONE MILLS

Ed Cole's original small-block wasn't really all that small in 1955, especially from the perspective of a driver fondly familiar with Chevrolet's proven 235-cid inline-six. Industry-leading displacements would soar soon enough, but 265 cubes were still considered a handful sixty years back, and similar opinions applied concerning the 327 V-8 when it debuted in 1962.

So what's in a name? Originally putting the small in small-block was its overall external dimensions, which were far more compact compared to most V-8s seen to that point. Nestled nice-and-tidy-like way down there between 1955 Bel Air fenders, the 265 Turbo Fire might've even reminded some imaginative cuss of a little mouse in its hole, thus supplying early inspiration for a nickname that later came into vogue after Chevrolet introduced its definitely large 348-cid W-series family in 1958. The famed 409's forerunner, the 348 also was the first of Chevy's so-called big-blocks. And when its successor, the Mk IV V-8 debuted, at 396 cubes in 1965, the nickname game evolved fully. In slang terms, the small-block became known as a "mouse motor," the Mk IV a "rat." Get it?

"Mighty mouse" is more like it, considering that Chevy's small-block simply refuses to back down. A major milestone in 1955, it now proudly stands alone—five generations and far more than 100 million built later—as the longest-running, most prolific powerplant ever. "There are many great engines throughout automotive history," said global chief engineer and small block program manager Jordan Lee in January 2016. "However, I challenge anyone to name [one] that has contributed more to western civilization in terms of numbers or popularity as the small block. Our legacy is already sixty-one years young and still going strong!"

Originally displacing 265 cubic inches in 1955, Chevrolet's first-gen small-block was bumped up to 283 cubes in 1957. *GM*

Five generations of Chevy small-block V-8s have come and gone during the last sixty-plus years, with the last four all introduced beneath Corvette hoods. *GM*

Variety, too, has been a small-block trademark, with displacements ranging from 262 cubic inches to 427, outputs from 110 horsepower to 638. Throw in countless technical updates here and there—two-bolt main bearing caps versus four-bolt, small crankshaft journals versus medium versus large, carburetors versus injectors, variable valve timing versus conventional lifters—and the limbs on this family tree fan out like a parachute.

Progressive design stages especially abounded during the 1960s and 1970s, with the aforementioned 327 morphing into the long-familiar 350 in 1967. Three years later came the largest Gen-I small-block, the 400, which, despite its not-small displacement, was never intended for big-bully applications. Nor was the yeoman 307 (built from 1968 to 1973), the short-lived, short-changed 262 (1975–76), or the economical, environment-friendly 267 (1979–81). Introduced for 1976, the 305 V-8 tried to look tough at times but for the most part was more or less a mere shadow of former hot-to-trot small-blocks. Chevrolet's 302, on the other hand, was a raucous rodent bolted up for one purpose only—to transform Camaro into the wildest pony this side of the Pecos. For the tale of this smokin' small-block, offered only for Z28 from 1967 to 1969, turn to Chapter 4.

Pioneering computer controls emerged d[illegible]ng the Eighties to save the small-block's bacon and laid the groundwork for the first Gen-II, the reincarnated LT1, which showed up in 1992 to initially power the fourth-generation (C4) Corvette. Dropping the LT1 beneath Z28 hoods the following year allowed Camaro drivers to force-feed dust to their Mustang counterparts like it was candy. The battle for pony car supremacy escalated further in 1998, when the Gen-III LS1, introduced along with 1997's C5 Corvette, pushed the Z28 atop the 300-horsepower plateau.

In 1999, Silverado pickup (and GMC Sierra) buyers were presented with three optional Vortec V-8s based on the Gen-III design: the 4.8-liter LR4, 5.3-liter LM7, and 6.0-liter LQ4. Six years later, the 400-horsepower LS2 Gen-IV V-8 debuted along with the C6 Corvette, and a 390-horsepower LS2 also went into 2005's SSR sport truck. Another Gen-IV variation, the 5.3-liter LH6, carried over into Vortec ranks that year as an option for Trailblazer EXT, along with GMC's Envoy XL and Envoy XUV.

Producing three hundred horsepower, Trailblazer's Vortec 5300 V-8 featured cost-conscious "Displacement on Demand" operation. In Chevrolet's own words, "DOD has

Announced in April 2003 (for 2005's GMC Envoy and Chevrolet Trailblazer), GM's Displacement on Demand technology deactivated cylinders during light-load conditions to improve fuel economy. Now known as Active Fuel Management, this system shuts down half the cylinders in Gen-V small-blocks by switching off oil pressure to their hydraulic lifters. *GM*

the capability of disabling the combustion process in half the [LH6's] cylinders in certain driving conditions, enabling fuel savings of 6 to 8 percent." According to engineering chief Sam Winegarden, "technology such as Displacement on Demand demonstrates [the small-block's] adaptability in the face of evolving marketplace expectations." The LS2 did not incorporate DOD.

The tag soon changed to "Active Fuel Management," an innovation that was extended into V-6 ranks for 2007's Impala. But it was on the V-8 truck side of the fence that AFM tools did their best work early on.

"The latest [AFM small-blocks] are ideally suited for Silverado and the tough jobs that pickup customers do every day," announced Jordan Lee in November 2014. "We give customers the power and torque of a full-size truck engine when they need it, then use Active Fuel Management and other advanced technologies to seamlessly shift to four-cylinder operation when

Chevrolet revived the LT1 small-block, first seen in 1970, sans hyphen for 1992's Corvette, kicking off the engine's second generation in the process. The reborn LT1 then became standard for 1993's next-gen Z28. *GM*

MILESTONE MILLS (continued)

the truck is under lights loads, making the most of fuel efficiency."

While the fifth-generation Camaro's LS3 Gen-IV V-8 didn't feature Active Fuel Management in 2010, its L99 running mate did, and both AFM and variable valve timing were included in the C7 Corvette's Gen-V small-block package four years later. When the Gen-V LT1 carried over for the new sixth-gen Camaro in 2016, AFM was included in auto-trans applications.

So what possibly can come next for this tireless traveler?

"We're always looking forward," added Lee in 2016. "The Small Block team is the best it's ever been in terms of intellect, capability, development tools at our disposal, and a desire and passion to preserve the legacy. [This] engine family has a very bright future. The new stuff we're working on is incredible, no other way to describe it."

Stay tuned. Same mouse time, same mouse channel.

Gen-I Small-Block V-8s, 1955–1979

Year	RPO	CID	Bore & Stroke	Horsepower	Torque	Compression
1955	n/a[1]	265	3.75 x 3.00	162 at 4,400 rpm	257 at 2,200 rpm	8:1
1957	n/a[1]	283	3.87 x 3.00	185 at 4,600 rpm	275 at 2,400 rpm	8.5:1
1962	300	327	4.00 x 3.25	250 at 4,400 rpm	350 at 2,800 rpm	10.5:1
1967	L48[2]	350	4.00 x 3.48	295 at 4,800 rpm	380 at 3,200 rpm	10.25:1
1967	Z28[3]	302	4.00 x 3.00	290 at 5,800 rpm	290 at 4,200 rpm	11:1
1968	n/a[4]	307	3.87 x 3.25	200 at 4,600 rpm	300 at 2,400 rpm	9:1
1970	LF6	400	4.125 x 3.75	265 at 4,400 rpm	400 at 2,400 rpm	9:1
1975	L99	262	3.67 x 3.10	110 at 3,600 rpm	200 at 2,000 rpm	8.5:1
1976	LG3	305	3.736 x 3.48	140 at 3,800 rpm	245 at 2,000 rpm	8.5:1
1979	L39	267	3.50 x 3.48	125 at 3.800 rpm	215 at 2,400 rpm	8.2:1

***NOTE:** CID is cubic inch displacement; bore & stroke in inches*
[1]Base engine for V-8 models [2]Camaro SS only [3]Camaro Z28 only
[4]Base V-8 for full-size Chevys, Chevelle & Nova

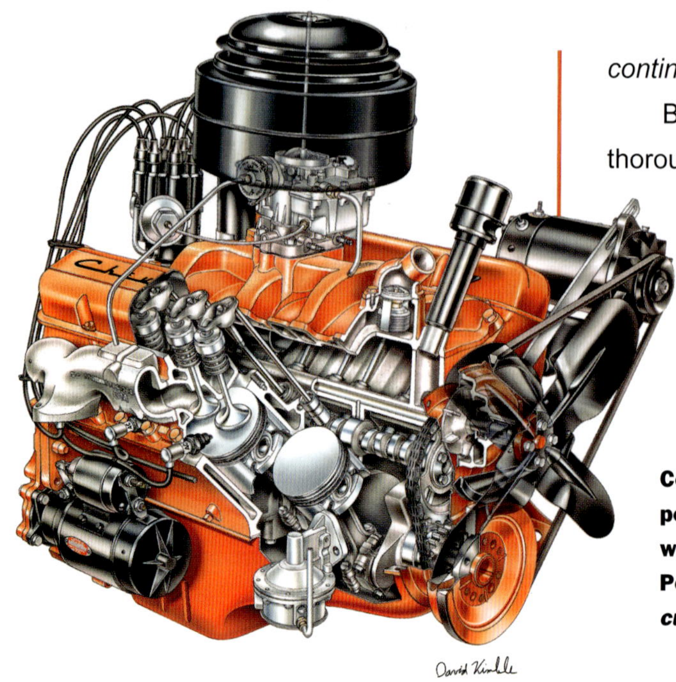

continued from page 21

But nothing in the cost-conscious compact arena could run with Iacocca's thoroughbred when 1966 began, not even Plymouth's Valiant-based Barracuda, the glass-backed rival that actually burst out of the gates first in April 1964. Same for Monza (the "fun" Corvair introduced in April 1960) and Chevy II's Super Sport alter ego, unveiled for 1963. Any hopes of disrupting this one-horse race hinged on GM fostering its own new breed. It was time to saddle up or stay on the porch.

Done and done.

Compact, lightweight, and powerful, Chevrolet's original small-block put real performance into the hands of Average Joe and friends in 1955. Its short stroke, working in concert with those ball-stud rocker arms (an idea copped from Pontiac), made 1955's 265-cid V-8 a high-winding wonderment. *David Kimble cutaway, courtesy GM*

Chevrolet's venerable small-block V-8 marked its sixtieth birthday in 2015. Even at such an advanced age, it remains a force both on the street and at the track. *GM*

Just like Mustang, 1967's all-new Camaro was meant to represent various different cars to various different drivers, with a thrifty six-cylinder starting things off in basic form. But its own flair for fun was still a prime attraction, a fact Estes (now managing Chevrolet) was understandably eager to promote. "Camaro is aimed at the fast-growing personal sports-type market that was pioneered by Corvette in 1953 and further defined by Corvair Monza in the 1960s," he said while introducing what he further called a "four-passenger package of excitement."

Helping build the excitement was the Super Sport rendition (a no-brainer in keeping with Chevy tradition) and the Rally Sport package, a cosmetic option that added trendy hideaway headlights. Additional choices included the SS 396, a big-block bully that appeared in November 1966 and acted more like a Clydesdale than a little horse. A few weeks later, the first Z28 began turning on even more buyers, and, by New Year's Day 1967, it was mostly Mustang owners bringing up the rear.

"Right from the beginning, Camaro was the 'closest thing to a Vette yet,'" claimed auto writer and marketing man Tony Hossain in 2007 while honoring GM's decision to revive the legacy. "Terms like 'SS,' 'Z28,' and 'Rally Sport' meant one thing: a very cool Chevy."

Save for a brief rest from 1975 to midyear 1977 and a transformation into the "IROC-Z" from 1988 to 1990, the famed Z28 rolled on rock steady into the new millennium, one of only three performance machines to survive the changing times that

continued on page 29

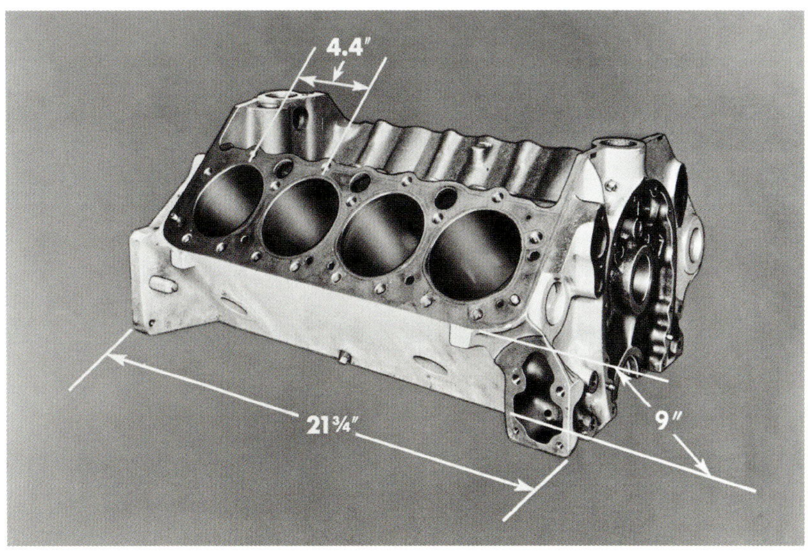

While so much has understandably changed during the small-block's long history, one feature remains constant: the 4.40-inch measurement between bore centers. *GM*

PACE SETTERS

When retired racer Dario Franchitti led the pack around the Brickyard in a Z28 coupe prior to the ninety-eighth running of the Indianapolis 500 on May 25, 2014, it marked the eighth time a Camaro had paced the field since this grand event originated in 1911. This also made it twenty-five momentous Memorial Day appearances for Chevrolet, dating back to 1948, when a Fleetmaster convertible was the first Bowtie-clad car to turn the trick. That total was a record, but one that didn't stand long: a Z06 Stingray made it twenty-six in May 2015. Camaro's count, meanwhile, ranks second only to Corvette's thirteen pace-lap performances.

A former Indy 500 champ (2007, 2010, and 2012), Franchitti was the fourteenth past winner to pilot a pace car on race day. He had announced he was walking away from open-wheel competition in November 2013, this after a last-lap crash during the Grand Prix of Houston the prior month left him with fractured vertebrae, a concussion, and some serious memory loss. But some guys just can't steer clear of the fast lane.

"It is a tremendous honor for me to be asked to drive the [Indy] pace car," he said in March 2014. "I will appreciate every minute of [pacing] the field in the new Z28. Although I won't be competing in the greatest spectacle in racing, this will be as close as one person can get to the action."

"Dario is a true champion who has earned the respect of competitors and race fans alike," added Chevrolet exec Jim Campbell while announcing Camaro's latest Indy assignment. "It will be very special to have [him] lead the field to the green flag in the Z28."

Like Ford's original Mustang in 1964, Chevy's first Camaro was invited to pace the Indy 500. And although not officially marketed as pace car replicas, about one-hundred look-alike models eventually found their way into private owners' hands in 1967 after serving complimentary duties prior to the race. Like the actual SS/RS pacer, these convertibles were all done in blue-striped white paint and included either 350 Turbo Fire small-blocks or 396 Turbo Jet big-blocks.

Three-time Indianapolis 500 winner Dario Franchitti paced the field of the 98th Indy 500 on May 25, 2014, in a Z28 coupe, making this the eighth time a Camaro has served as the prestigious pace car at The Brickyard. *GM*

The replica fad really got rolling two years later, when Chevrolet sold 3,675 Indy Pace Car Camaro SS/RS convertibles, all featuring Dover White paint complemented by orange houndstooth interiors. Rally wheels and cowl-induction hoods were included in all cases, too, and both small- and big-blocks again were installed. An actual RPO code, Z11, was assigned this time.

A similarly adorned pace car replica coupe—a promotional package created for Chevrolet's southwestern branch zone office—also appeared with many of the same features, listed under RPO Z10. No production numbers are available for the Z10 coupe, with estimates long claiming only 200 or 300 were released during the spring of 1969. But according to Camaro Research Group (www.camaros.org) tabulations, the production count was at least 350, if not between 400 and 500.

A third Pace Car replica followed in 1982, just in time to help mark the arrival of the new third-generation

Camaro. Chevy sold 6,326 this year, all Z28s featuring a silver/blue finish and complementary interior. And a fourth, again commemorating the arrival of a next-gen F-car, was released for 1993 to help put its parent company alone up on a pedestal at Indy for the first time.

"Records are meant to be broken, particularly when they're your own," said Chevrolet general manager Jim Perkins in 1993 while announcing the new fourth-gen pony car's upcoming race-day duties. "Until now, Chevrolet and our sister division Oldsmobile were tied at eight for most pace car appearances, and Camaro's three pace car titles tied it with the Stoddard-Dayton from the early 1900s."

Chevy's fourth Camaro Indy pacer also was the first F-car to hit the Bricks with no drivetrain or chassis modifications, a feat originally managed by Corvette in 1978. "Performance-wise, every 1993 Z28 is capable of pacing Indy [right off the truck]," continued Perkins. "With the LT1 V-8 engine, there's no need for additional power."

The same could be said for the all-new fifth-generation Camaro, which paced the Indianapolis 500 in 2009, 2010 and 2011. An SS coupe led the way that first year, but its pre-production status negated any chance of street-going copies. A limited run of flaming orange replica coupes then appeared in 2010, followed by a second batch of convertibles in 2011, all done in a familiar orange-striped white finish.

"It's an honor for Chevrolet to pace the Indianapolis 500, especially during this landmark year," said marketing vice president Rick Scheidt in February 2011. "This is

Along with pace lap appearances at Indianapolis in 1967, 1969, 1982, and 1993 (left to right in back), Camaro also paced the NASCAR field for 1996's Brickyard 400. The three Z28 coupes prepared for that race were 1996 models dressed up in 1997 30th Anniversary garb. *GM*

Camaros paced the Indianapolis 500 three years running beginning in 2009, with a convertible finally showing up in 2011. Like the actual pace-lappers, 2011's street-going replicas (right) honored 1969's Indy pacers, at left. *GM*

PACE SETTERS *(continued)*

also a great opportunity for us to build on Camaro's strong momentum, pay tribute to its history and celebrate the ongoing launch of the new convertible."

The landmark Scheidt spoke of was his company's one hundredth year in business.

Yet another centennial was marked in May 2016 with the one hundredth running of the Indianapolis 500. And just as these pages were going to press, Chevrolet announced that a Camaro would once again pace the Indy field, a 50th Anniversary coupe driven by living legend Roger Penske, who also celebrates a half century in the race team ownership business.

Indianapolis 500 Pace Car Chevrolets

Year	Model	Driver
1948	Fleetmaster convertible	Wilbur Shaw
1955	Bel Air convertible	Thomas Keating
1967	Camaro SS/RS convertible	Mauri Rose
1969	Camaro SS/RS convertible	Jim Rathmann
1978	Corvette coupe	Jim Rathmann
1982	Camaro Z28 coupe	Jim Rathmann
1986	Corvette convertible	Gen. Chuck Yeager
1990	Beretta convertible	Jim Perkins
1993	Camaro Z28 coupe	Jim Perkins
1995	Corvette convertible	Jim Perkins
1998	Corvette convertible	Parnelli Jones
1999	Monte Carlo coupe	Jay Leno
2002	Corvette convertible	Jim Caviezel
2003	SSR sport truck	Herb Fishel
2004	Corvette convertible	Morgan Freeman
2005	Corvette convertible	Gen. Colin Powell
2006	Corvette Z06 coupe	Lance Armstrong
2007	Corvette convertible	Patrick Dempsey
2008	Corvette Z06 coupe	Emerson Fittipaldi
2009	Camaro SS coupe	Josh Duhamel
2010	Camaro SS coupe	Robin Roberts
2011	Camaro SS convertible	A. J. Foyt
2012	Corvette ZR1 coupe	Guy Fieri
2013	Corvette coupe	Jim Harbaugh
2014	Camaro Z28 coupe	Dario Franchitti
2015	Corvette Z06 coupe	Jeff Gordon
2016	Camaro 50th coupe	Roger Penske

Camaro Pace Car Replicas

Year	Production
1967	see NOTE below
1969	3,675 Z11 convertibles[1]
1982	6,360 1FP87 coupes
1993	633 B5A coupes
2009	none built
2010	294 Z4Z coupes
2011	500 Z4Z convertibles
2014	none built[2]

NOTE: *GM paperwork shows 44 festival cars, 71 dealer specials, 7 replicas, 5 "unknown," and 21 Canadian*

[1]Estimates claim 350–500 Z10 coupes

[2]While no replicas were built, Chevrolet did offer three Indy Stripe options to accompany 2014's RS package: RPO DTC (black), DTA (gray), and DTE (silver). Production was 557 for DTC, 120 for DTA, and 317 for DTE.

continued from page 25

knocked all other muscle cars dead in their tracks for good during the Seventies. Pontiac's ever-popular Trans Am never stopped takin' it to the streets from 1969 to 2002, and Chevy's Corvette is still out there proving it all night in its sixty-third straight year.

On November 29, 2011, Chevrolet's small-block really did go where no engine has gone before as technicians, VIP journalists and key past and present team employees in Wixom, Michigan, put the finishing touches on the 100 millionth example of this long-running V-8. *GM*

Every bit a major part of the Camaro story is another undying Bowtie legacy, this one entering its sixth uninterrupted decade (and fifth generation) as these words head to press in 2016. Offered between pony car flanks along with that practical six-packer in the fall of 1966 was Chevrolet's ever-present small-block V-8, then running a dozen years into its initial Gen-I chapter. The markedly upgraded 5.7-liter Gen-II followed (first for Corvette only) in 1992, then was superseded by Chevy's radically redesigned Gen-III five years later, again as the exclusive heart of the latest, greatest plastic-bodied two-seater. Horsepower hounds knew the Gen-II and Gen-III V-8s best by their top-shelf labels: LT1 and LS1, respectively.

New for 2005, along with the first sixth-generation (C6) Corvette, was the LS2 Gen-IV, a 6.0-liter aluminum marvel that put out 400 horsepower. And when the C7 Stingray debuted for 2014, it featured the Gen-V, a 455-horse V-8 that revived that revered LT1 tag.

"The [latest] LT1 represents the most significant redesign in the small-block's nearly 60-year history—building on its legacy to make one of the world's best engines even better," beamed Global Powertrain Engineering vice president Sam Winegarden while introducing the C7's heart and soul in October 2012. "[It] is a sweetheart of a powerplant, and drivers will feel its tremendous torque and power at every notch on the speedometer."

But these were just high-performance highlights. Each generation also fostered other less prominent versions for everyday transports, and it was these powerplants that helped put Chevrolet's so-called "mouse motor" into the record books. In a world newly crowded with ultra-high-tech overhead-cam motors, Chevrolet's brilliantly-engineered small-block V-8 remained one of the most potent engines on the planet, pushrods and all. In November 2011, Chevy released its 100 millionth small-block, by far the highest production plateau ever reached by any engine family anywhere.

"The critics can say what they want about the relevance of pushrod engines," said *Wards Auto World* executive editor Tom Murphy in December 2013. "But how can more than a hundred million buyers of small-block engines since 1955 be wrong?"

A provisional Trans-Am competitor fifty years ago, Chuck Bently's '67 Camaro still competes in Sports Car Club of America vintage racing today, touring Sebring's treacherous turn 7 hairpin here. ***Tom Shaw***

Pushrods were all the rage around Detroit when Chevrolet introduced its original overhead-valve small-block, labeled the "Hot One" by promotional people. GM kicked off the postwar OHV V-8 craze in 1949, first via Cadillac, then Oldsmobile immediately thereafter. Chrysler joined the club with its Firepower V-8 in 1951, and this innovative "hemi-head" design was then passed down to DeSoto in 1952 and Dodge in 1953. Indiana-based Studebaker became the first independent to offer a modern OHV V-8 in 1951, with Lincoln the following year, followed by Buick in 1953 and Mercury in 1954. Packard, Hudson, Plymouth, and Pontiac all made the grade the next year.

As for "U.S. #1," this country's long-standing sales leader, Chevrolet began developing its ground-breaking OHV Turbo Fire V-8 in December 1951. Plain six-cylinder practicality had long been the main key to Bowtie dominance, but the annual runner-up, Ford, was modernizing fast. The word wasted little time getting around about Dearborn's work on a new OHV V-8, for 1954, to replace its obsolete side-valve "flathead," which had been powering Blue Oval products since 1932.

Not to be left resting on their laurels, GM execs in April 1952 moved Ed Cole and Harry Barr over from Cadillac, Cole to serve as Chevrolet chief engineer. Both had played major roles in Cadillac's short-stroke, OHV V-8 design process, and now they were tasked with working similar magic for the luxury line's lower-priced corporate comrade.

The results were historic, to say the least. Triple digits on the speedometer had never before come so cheap, and dropping the Turbo Fire between fiberglass fenders in 1955 surely saved Corvette from an early death.

"The performance of the small-block transformed Chevrolet," said GM Performance Vehicles and Motorsports vice president Jim Campbell in August 2011. "[It] made Chevrolet the weapon of choice for grassroots racers on drag-racing and sports-car tracks across America. It also powered Chevrolet's factory racing programs, leading to wins in stock car, endurance, and Trans-Am

Camaro Websites

	Organization name	Description
www.crg.com	Camaro Research Group	By far the best source out there for 1st-generation info
www.camarotech.com	Camaro Tech	Focuses on everything 4th-generation
www.chevy-camaro.com	Chevy-Camaro.com	Mucho 1st- through 4th-generaton material
www.nastyz28.cm	Second Generation Camaro Owners Group	As the name implies, 2nd-gen cars are the focus here
www.35thle.com	35th Limited Edition	Dedicated to 2002 Anniversary Camaro SS, but also features some other neat info
www.thecamaro.com	theCamaro.com	Deals with almost all things Camaro
www.moderncamaro.com	ModernCamaro.com	Deals with the latest, greatest news
www.firstgens.com	FirstGens.com	Features a nice forum for 1967-1969 people
www.camaro5.com	Camaro5	5th-gen fun

RPO CODES

In 1963, Chevrolet Motor Division officials revised their options coding system, dropping the three-digit nomenclature used previously in favor of an alphanumeric Regular Production Option (RPO) list that cleanly broke down the seemingly countless codes into related groups. For example, all Chevy engines were tagged with an "L" prefix, all transmissions with an "M." The really good stuff was found in the "Z" category, labeled "Chevrolet Special Items." Z11, Z16, Z28—you get the picture. Changed and rearranged a few times over the years (various group counts expanded, and lengthened codes such as"LS6" and "ZR1" appeared), the original RPO structure looked like this a half century back:

Code	Option Group	Code	Option Group
A01 through D99	Body	N30 to N49	Steering
F01 to F39	Frame	P01 through S99	Wheels & tires
F40 through F99	Front & rear suspension	T50 to T59	Sheet metal
G50 through H99	Rear axle	T60 through U99	Electrical & instruments
J50 to J99	Brakes	V01 to V29	Radiator and grille
K01 through L99	Engine	V30 to V99	Bumpers & miscellaneous
M01 to M99	Transmission	Z01 to Z49	Chevrolet Special
N01 to N29	Fuel & exhaust		

series. [That] racing experience in turn led to more potent production cars, creating legendary names like Corvette, Camaro, Impala, and Chevelle."

Unprecedented compactness also worked in the first small-block's favor. At 530 pounds, it measured 42 less pegs than the venerable six it overshadowed. As for power, advertised output for Chevy's passenger-car "Stovebolt" in 1955 was 136 horses, compared to 162 for the new base V-8. Less mass, fewer pounds, yet more displacement, thus increased output—was this magic? Not at all, it was simply GM engineering supremacy struttin' its stuff.

Back in 1955, it was case closed as soon as Ed Cole's first small-block hit the streets: this was *the* greatest thing since sliced bread. And it's still slick as melted butter sixty years later. As in 1993 and 1997, the latest next-gen small-block rolled over from Corvette ranks into the Camaro clan for 2016, catapulting Chevrolet's all-new sixth-generation pony car to all-time heights with 455 honest horses.

"We have had the incredible opportunity to meet literally thousands of Gen 5 owners who provided direct feedback on what they loved about their car and what they wanted for the next-gen Camaro," said chief engineer Al Oppenheiser in May 2016. "As a result, the 2016 Camaro builds on what made the current [model] such a success with more power, more agile handling and more technology. We expect it will set the new benchmark in the segment—and give a new generation of enthusiasts a reason to fall in love with Camaro."

May death alone do them part.

CORVETTE
CHEVY II
CHEVELLE
CHEVROLET
CAMARO

HISTORY IN THE MAKING

Chevrolet brings its F-car to market

Funny how Ford gets every column inch of credit for inventing Detroit's pony car, especially considering that Chevrolet's Monza-ized Corvair had helped inspire Lee Iacocca's comparatively bigger baby. Granted, after hitting the ground in a gallop in 1964, Mustang did help redefine the way average young Americans looked at everyday transport. Ah, but hadn't the affordable, innovative Monza already managed that leap four years before?

"Anyone interested in a different kind of car should have a ride in a Monza," announced *Automobile Topics'* Floyd Clymer in 1960. "It will change their conception of motoring, for this is a fun-to-drive [automobile]."

"It has all the appearance of a Gran Turismo machine," echoed *Sports Car Graphic*'s Jerry Titus in 1961. "It's even more fun than before." Inspiring Titus's comparison was that year's new option, the 98-horsepower, 145-cid Super Turbo Air engine, a "pancake" six that was 18 horses stronger than Corvair's standard, mild-mannered mill.

Fanning the flames of frivolity further in 1962 were 102 optional horses and an even more titillating Monza, the Spyder, called "a poorman's Porsche" by *Car and Driver*. Spyder venom was supplied by a turbocharged six that made 150 horsepower up through 1964, 180 in 1965. Handling those horses, however, was something many Americans—at the time mostly unaccustomed to early turbo travel's trials/tribulations—couldn't quite grasp. According to *Automobile*

(opposite)
Camaro (foreground) made it six model lines for Chevrolet in 1967, joining (counterclockwise above) Chevelle, Corvette, Chevy II, and Impala. Absent here is Corvair, by then rolling well on its way out of the Bowtie picture. *GM*

Quarterly contributor Karl Ludvigsen, this was "the first modern engine built and sold by Detroit that required a skilled, interested driver to extract the best performance from."

Less interested auto-pilots simply enjoyed Monza, boosted or not, for what it was: a budget buggy with flair. Standard stuff included extra trim and bucket seats, youth-oriented teasers Iacocca unabashedly knocked off when breeding time came for his little horse. So why did the intriguing Corvair—with its independent suspension at all four corners and air-cooled, optionally turbocharged engine out back—roll off into the archives while the comparatively mundane Mustang made history?

Various reasons, for sure; but the main problem involved the simple fact that the Mustang was more in tune with the desires of America's car-buying public, in spite of its yeoman Falcon roots–perhaps even *because* of its pedestrian engineering. The Corvair was possibly *too* innovative for its time, offering the public advanced technology that it was not ready to

FROM THE HORSE'S MOUTH II

"Chevrolet is definitely going to have a new car included in the new models for the 1967 model year, and this car will be introduced on September 29, 1966. This is not news to many of you. As far back as April 8, 1965, a Detroit newspaper had page-one headlines proclaiming that Chevrolet had a 'Panther' car to answer the horsy offering of a competitor. The question of 'who was answering who?' has been discussed frequently, since [that] four-footed competitor really answered the Corvair. And the Corvair actually followed up the Corvette, which was the real pioneer of the sporty-personalized car in this country. Much as we appreciate the tremendous publicity given 'Panther,' we ask you to help scratch the cat once and forever. Chevrolet has chosen a name which is lithe, graceful, and in keeping with our other car names beginning with the letter 'C.'"

Chevrolet General Manager Pete Estes during a June 29, 1966, press conference, from archival notes researched by Camaro historian and former assistant brand manager Scott Settlemire

Former Pontiac chief Pete Estes moved over to Chevrolet's general manager chair in July 1965, just in time to hopefully help blow away the competition in Detroit's burgeoning pony car race. *GM*

Chevrolet's innovative air-cooled Corvair was mostly an affordable compact when introduced for 1960. Then along came the sporty Spyder, with its turbocharged pancake-six, two years later. Available with or without a top, the 150-horsepower Spyder was offered up through 1964. Pictured is a 1963 model. *GM*

embrace. The end result was that Ford sold more Mustangs than Chevrolet sold Corvairs.

Although General Motors had invested much time, energy, and capital in developing the technologically advanced Corvair, it soon became clear that Chevrolet needed to go in a more traditional direction.

Initially General Motors placed its bets squarely on an exciting redesign then in the works for the 1965 Corvair. As it turned out, this beautiful body design, while a real head-turner, lost all chance of making hay in the compact marketplace after Ralph Nader's book *Unsafe at Any Speed* hit shelves in late 1965, but by then the decision had been made to field a competitive front-engine, rear-wheel-drive coupe.

On Friday, April 17, 1964, Dearborn folk initiated a vehicle launch the likes of which industry watchers had never before seen. Showroom cars were claimed even before curtains lifted, another twenty-two thousand were ordered that first day, and as many as four million potential buyers rushed their neighborhood Ford lots that weekend to see what Iacocca had wrought.

CAMARO CORRALS

Chevrolet built Camaros at two manufacturing locations from 1967 to 1971: one in Norwood, Ohio, the other in Van Nuys, California. The Norwood plant rolled out Chevy's pony car on its own from 1972 to 1975, then production restarted at Van Nuys in February 1976. Both sites shared the load again until Norwood closed in 1987, leaving its SoCal counterpart this time to carry on alone up through 1992. Van Nuys was shuttered after releasing its final F-body (on August 28 that year) and later demolished. Production then moved north of the border to Sainte-Therese, Quebec, when the fourth-gen model hit the ground running in January 1993, and this thirty-seven-year-old facility also shut down forever (in August 2002) after completing its last Camaro. Oshawa Assembly, in Ontario, Canada—one of the world's largest and most efficient automotive manufacturing ventures—was home to fifth-generation models during their short, happy six-year run. Today, Chevrolet's sixth-gen Camaros are built near Detroit at GM's second-youngest facility back here in America, the Lansing Grand River Assembly plant. Opened in 2001, Lansing Grand River also produces Cadillac's CTS and ATS.

In 1963, Buick targeted Ford's Thunderbird with its lovely Riviera, proving yet again that a sporty coupe could sell, even at luxury liner price levels. Some at GM, including Chevrolet design studio head Irv Rybicki, then concluded that a more affordable take on this theme surely would bring even more buyers into Chevy showrooms. *GM*

Mustang's sales reached six digits after only four months. In February 1966, Ford's pony became the quickest production car ever to reach seven-digit sales numbers. More than 1.3 million Mustangs were prancing about by the time Chevrolet's direct response made the starting field later that year.

Not that Chevrolet had not explored sporty options. Consider the "Super Nova," an exciting concept car conceived in Hank Haga's Chevrolet Number Two Studio late in 1963. Chief designer Irv Rybicki's baby, this small, sporty "personal coupe" was inspired by Buick's new-for-1963 Riviera, which in turn was a response to Ford's established Thunderbird, the cruiser that originated Detroit's "personal luxury" niche. Rybicki's downsized dream machine, initially created in supreme secrecy, was based on the conservative Chevy II's platform, but its low lines and exceptionally clean looks invited liberal amounts of oohs and ahs during its unveiling at the New York Auto Show in April 1964. Witnesses also marveled at its "T-birdesque" long hood and short rear deck, which provided a profile further familiarized by Mustang just two weeks later during its historic Big Apple intro at the 1964 World's Fair.

Chevrolet general manager Bunkie Knudsen was all in favor of rushing this lively two-door coupe to market, as was Bill Mitchell and his predecessor, retired GM Styling mogul Harley Earl. "[Super Nova] was an idea car, but not so exotic that you couldn't produce it," explained Knudsen to Gary Witzenburg fifteen years later. "I recall some Ford people asking whether we were going to build [it], and I remarked, 'if they'll let us, we'll build it.'"

GM execs weren't keen to add a sixth model to Chevrolet's already full lineup when Irv Rybicki's sensational Super Nova concept debuted at the New York Auto Show in April 1964, beating Mustang onto the scene by two weeks. Chevy general manager Bunkie Knudsen had hopes for this clean machine making it into production, but his bosses said no. *GM*

But they didn't. Corporate president Jack Gordon put the kibosh on this promising proposal. However, Gordon and crew changed their minds after witnessing the record-breaking sales of Ford's Mustang, and on August 8, 1964, Knudsen received the green light for a real response: Called "F-car," this rapid-fire response was scheduled to go on sale by the fall of 1966, an extremely tight deadline. Haga's studio wasted little time sculpting a prototype Camaro, code-named XP-836.

Initial engineering specs were completed by August 25, 1966, and early test mules were up and running the following month. Most of the final look showed up in December in the form of a full-size clay that followed in Mustang's hoof

TWO-HORSE RACE

Chevrolet people have equated this epic confrontation to the classic Hatfield/McCoy family feud, but without all the blood and bad feelings. "For five decades, Camaro and Mustang have been battling it out in every possible setting," said GM Global Product exec Mark Reuss in April 15, 2014, two days before Ford's pony celebrated its 50th birthday. "These two cars have been striving to beat each other on the track, on the drag strip, and on the street. That competition is a big part of why both cars are so amazing, and so popular, today." "The new Camaro hit the ground running [in 1967], and it's been a battle royale ever since with the Bowtie vs. the Blue Oval," added veteran auto writer Ken Gross. "Chevy readily admits its Camaro is constantly getting better, thanks to intense competition from Ford, and I'd agree. Competition *always* improves the breed. I'll bet these two legendary nameplates will still be competing 50 years from now."

Ford's pioneering pony car celebrated its 50th birthday with two gala celebrations, one in Charlotte, North Carolina, the other in Las Vegas, in April 2014. Although they couldn't come (or was it that they weren't invited?), Chevrolet promotional people at least sent a card. *GM*

prints in nearly all aspects. The hood was lengthy, the rear deck less so, and overall impressions were truly sporty. What set the F-car apart was its softly contoured shape, which instantly made Ford's original pony look stiff and boxy in comparison. As for basic dimensions, the new challenger was slightly longer, lower and wider, and predictably featured a little more passenger room inside.

The first hand-built prototype was completed on December 16, 1965, and pilot assembly started on May 17, 1966. Regular production commenced on August 7, 1966, then Chevy's first Camaros hit dealer showrooms on September 29, 1966. *Rush* to market? Rampage was more like it.

LEADING MAN: ED COLE

Ed Cole knew a thing or two about building V-8s when he moved over from Cadillac to Chevrolet in 1952, and one of the main lessons learned involved durability. "An engine must be made to hang together under any circumstances," he later said. That the small-block Chevy has hung on strong now for sixty years certainly stands as an undeniable testament to his expertise. Born in Marne, Michigan, in 1909, Cole was demonstrating mechanical proficiencies at an early age. Although he initially enrolled at Grand Rapids Junior College with intentions of eventually earning a law degree, he soon transferred to the General Motors Institute for engineering training. After joining a work/study program at Cadillac, he was hired on by GM's flagship division as a full-time engineer in 1933. Among early feathers in Cole's cap were various military design projects, including a new engine for the US Army's M-3 light tank. Next came the assignment that made his career: Cadillac's OHV V-8, introduced for 1949. Within a few years he was managing Cadillac's Cleveland plant, then was transferred to Chevrolet, where he helped make history in 1955. Cole was rewarded in July 1956 with a promotion to Chevrolet general manager and GM vice president, then on to president of General Motors in 1967. He retired in 1974. He died tragically three years later in a private plane crash.

After making history with Chevrolet's "Hot One" in 1955, GM engineering legend Ed Cole's next milestone moment came late in 1959, when he introduced the rear-engine Corvair to America's then-formative compact crowd. *GM*

The familiar pony car profile—long hood, short rear deck—certainly was nothing new when Mustang made the scene in 1964. Many witnesses then were quick to compare Ford's pioneering sport compact to another beloved Blue Oval product, the original two-seat Thunderbird, a modern classic built from 1955 to '57.

Beneath that attractive, mildly curvaceous skin was unibody construction, same as Mustang but with additional notable differences. Whereas Mustang was fully unitized with stamped-steel sections welded onto the body up front to carry the engine and suspension components, Chevrolet's newcomer featured a strong ladder-type front subframe that bolted up to the underbody. To help hit the short deadline, F-car project engineers simply relied on much of the same platform already being readied for 1968's next-generation Chevy II.

"Chevrolet engineers picked this half-unitized, half-frame construction for very good reasons," explained a *Car Craft* review in 1968. "Conventional unit-body/chassis structures tend to transmit a good deal of noise and vibration into the driving compartment. Yet the older style body-mounted-on-a-full-frame system has the disadvantage of ponderous weight and bulkiness. So by copying the half-and-half plan used so successfully in the Chevy II, the Camaro's power unit could have its own isolated frame while passengers ride in a lightweight, rattle-free cockpit behind it."

Chevy II's Mono Plate springs also carried over in back. But in the F-car's case, those single leafs were shorter and not parallel—their rear mounting locations were moved outward to make ample room for an 18-gallon

Initially tagged "Panther," Chevrolet's XP-836 project got rolling in August 1964. Soon-to-be-familiar Camaro cues were already showing up in clay by Thanksgiving that year. *GM*

LEADING MAN: ELLIOTT "PETE" ESTES

Born January 7, 1916, Pete Estes earned his engineering chops under GM legend Charles Kettering and contributed to Oldsmobile's groundbreaking high-compression Rocket V-8, introduced in late 1948. He initially joined GM's Research Laboratories in 1934 while a student the General Motors Institute in Flint, Michigan, and followed up his GMI studies with a two-year stint at the University of Cincinnati, earning a mechanical engineering degree in 1940. He took a development engineer position at Olds in 1946 then moved over to the chief engineer's seat at Pontiac ten years later. Pontiac's general manager position became his in November 1961, as did the same job at Chevrolet in July 1965. In February 1969, Estes was promoted up to head GM's car and truck group, then was named executive in charge of Overseas Operations a year later. His title became GM director and executive vice president in charge of Operations Staff in October 1972, and he reached GM's top office in September 1974. He then served as CEO and president until his retirement on February 1, 1981. Called "one of GM's brightest stars," by chairman Robert Smith, Pete Estes was seventy-two when he succumbed to a heart attack on March 24, 1988. "In whatever role he assumed for the corporation, he performed at peak capacity," said Smith the day after Estes's death. "He demanded excellence from his subordinates, but he always drove himself harder than anyone else."

Pete Estes may have had more fun during the 1960s than any other Detroit executive. After shepherding Pontiac's GTO into production in 1964, he did the same for Camaro three years later. *GM*

gas tank. And Camaro's front suspension was conventional (like Chevelle's) with a coil spring stuck between unequal A-arms on each side. Similar to Ford's Falcon (and, of course, its Mustang derivative), the Chevy II's front suspension featured coils mounted *atop* the upper control arms.

The F-car's rear axle also came from Chevelle, not Chevy II, and adding this component widened the track originally laid out by engineers. This was by no means a bad thing, especially from where the designers were sitting—that extra width translated into a more aggressive look. And wouldn't you know it? As it always does, more track also translated into enhanced roadability.

"Handling actually came out better than we'd intended," explained Chevrolet engineering chief Alex Mair to *Car Craft* late in 1967. "The car's directional stability, particularly, is better than predicted. And there are

several reason for this. One is that we widened Camaro's tread during its late development stages—set the rear wheels father apart. That caused some gain in handling capabilities."

Computers, too, helped Camaro handle better than projected, making this Chevrolet's strongest commitment yet to this still-new technology—if only because, as Alex Mair told author Mike Lamm in 1978, "we didn't have enough time to do the old seat-of-the-pants work." "I think we used more computer programming on that particular project than [on] any car we'd done previously," added chassis engineer Edward Nash.

Looking very much like the final product, an XP-836 model was compared to its established rival in December 1964. ***GM***

Camaro benefitted both from pioneering computer design inputs and aerodynamic wind tunnel testing. The latter work was performed in Texas on a quarter-scale model by stylist John Schinella (left) and engineer Charles Rubly (in vest) in November 1965. ***GM***

As chief drivetrain engineer Don McPherson explained in a 1967 promotional film, space-age technology was used extensively to evaluate "such basic information as the height of the car, its weight distribution, center of gravity and wheelbase dimensions. The computer comes up with the best combination of such factors as roll axis, roll distribution, roll stiffness, ride rate, front-end geometry and so on."

The Camaro design process also benefited from another technology then taking root in the automotive world: wind-tunnel testing. In February 1965, a quarter-scale F-car coupe model was sent to the Ling-Temco-Vought aeronautics facility outside Dallas, Texas, for eleven days of wind-tunnel

LEADING MAN: IRV RYBICKI

On August 1, 1977, Irvin Rybicki became only the third man ever to head GM's Design department, following in the Godzilla-sized footprints of William Mitchell and Harley Earl. Born September 16, 1921, in Detroit, Rybicki began sketching and modeling cars at an early age and received art training while in high school. After serving in the army during World War II, he took a GM Proving Ground project engineer position in February 1944, then transferred to GM Styling Staff (renamed "Design Staff" in 1956) in September 1945. A senior designer by 1950, he spent six years in the Cadillac studio under Mitchell's direction before being named assistant chief designer at Oldsmobile. He was promoted to Olds studio head the next year, then moved over to the same office at Chevrolet in February 1962. In July 1965, Rybicki earned a promotion to chief designer for both Chevrolet passenger cars and GMC trucks, and in September 1970 was promoted again to executive in charge of exterior design for Chevrolet/Pontiac cars and Chevy/GMC trucks. He was put in charge of Olds, Buick, and Cadillac designs in September 1972 and was elected GM Design's vice president following Mitchell's retirement five years later. Chuck Jordan then became Design's fourth vice president following Rybicki's retirement in October 1986. He died in July 2001.

Even before Ford proved the concept doable in a really big way, Irv Rybicki was sure his sporty compact idea would fly. But his superiors didn't agree until Mustang made them true believers. In 1977, Rybicki became the third man to head GM's Design department, following in the formidable footsteps of Harley Earl and Bill Mitchell. *GM*

experiments, which resulted in only a few minor updates. Apparently that aggressive-looking shell also was inherently aerodynamic, probably more of a lucky break than a product of technical experience.

Meanwhile, designers back in Michigan were playing around with other body shapes, including a truly sporty two-seat convertible and two-door station wagon, called "Kammback." Though certainly appealing, this pair had no chance of ever finding the necessary funding (not to mention approval) to go from dream stage to the real world. The same reality applied to a fastback rendition, which would've matched up against Mustang's third body style. From the get-go, the plan only involved two Camaros: sport coupe and convertible.

As for its name, company insiders first called Chevy's F-car "Panther," a moniker also picked up by the press. On April 8, 1965, *Detroit News* reporter

Yet another in a long line of concept cars built by order of GM Design head Bill Mitchell, 1967's Cherokee convertible began life on the Norwood, Ohio, line as a 396-equipped RS/SS. Along with custom paint, modifications included a fiberglass hood fitted with a transparent scoop. Also notice the Corvette-style split bumpers. *Tom Shaw*

Tale of the Tape: 1967						
	Wheelbase (inches)	Weight	Base Price	Base 6-cyl.	Base V-8	Production
Chevrolet Camaro	108	2,770, 6-cyl. 2,920, V-8	$2,466, 6-cyl. $2,572, V-8	230-cid inline	327-cid	220,906
Pontiac Firebird	108	2,955, 6-cyl. 3,323, V-8	$2,666, 6-cyl. $2,761, V-8	230-cid inline OHC	326-cid	82,560
Ford Mustang	108	2,696, 6-cyl. 2,885, V-8	$2,461, 6-cyl. $2,567, V-8	200-cid inline	289-cid	471,121
Mercury Cougar	111	2,851, V-8[1]	$3,005	n/a	289-cid	150,893
Plymouth Barracuda	108	2,730, 6-cyl. 2,855, V-8	$2,449, 6-cyl. $2,530, V-8	225-cid slant-six	273-cid	62,534

NOTE: *Data shown here for comparable two-door coupe models; production count is the total for all models; weight in pounds*
[1]No six-cylinder offered

Jack Crellin wrote that Chevrolet was rapidly preparing a "Mustang-type" car allegedly based on the Super Nova show car. Along with predictions that Pontiac and Buick would offer versions of this pony-beater, Crellin's article also mentioned potential names: "Chevette" and "Colt," along with "Panther." Later, *Popular Science*'s December 1965 issue featured a preview, written by auto editor Jan Norbye, that announced that "Chevy is getting ready to pitch its Panther against Ford's Mustang in Detroit's biggest dogfight since they threw the 'stovebolt six' against old Henry's Model A."

But Panther struck out. Or was it shot down? Either way, GM bailed on this dangerous-sounding name after Ralph Nader and his "Raiders" began calling automakers to task for marketing their wares in such aggressive, irresponsible fashion. Besides, Panther didn't begin with *that* letter.

Reportedly, Chevrolet employees made about five thousand suggestions before Bob Lund and GM Car and Truck Group vice-president Ed Rollert put their heads together and came up with "Camaro," a traditionally correct

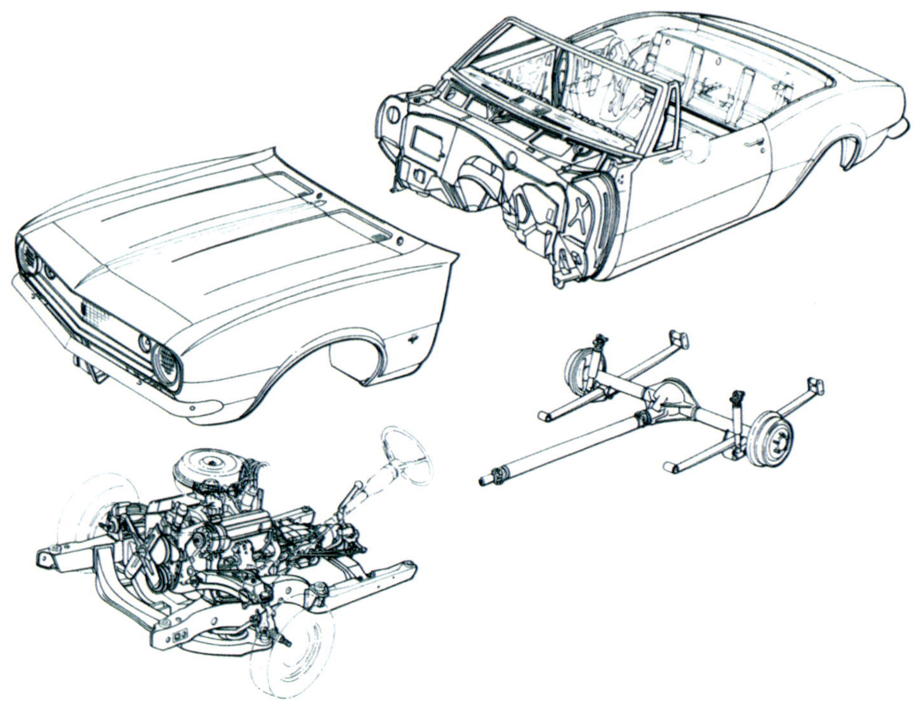

Chevrolet's F-body platform technically qualified as "semi-unitized," in that its main body structure was bolted up to an isolated front subframe. Mustang was fully unitized; that is, a front "apron" was welded up to its body. Notice the Chevy II–sourced Mono Plate single-leaf springs in back. *GM*

(alphabetically speaking) noun that, in Lund's opinion, had "kind of a ring" to it. In French it means "friend," "comrade," or "pal." But in Spanish it refers to a "shrimp-like creature." Guess which image Chevrolet was counting on.

When Chevrolet general manager Pete Estes (who had taken over from corporate-ladder-climbing Knudsen in July 1965) officially announced Camaro during a press conference in June 1966, he allowed a group of six Michigan State University cheerleaders, holding large block letters, to spell things out. Or so he hoped. Their first try resulted in "CAMAOR" after the last two coeds apparently took wrong turns on the way into the conference room. Once corrected, the sweater-clad lineup still left many wondering "what the hey?," at least until they were all given English-French dictionaries and encouraged to ignore the Spanish translation. Estes went on to explain how the French rendition fit because "the real mission of our new automobile [is] to be a close companion to its owner, tailored to reflect his or her individual tastes."

Fifty years later, that friendship stands stronger than ever.

Wearing VIN 100001, this six-cylinder coupe was the first of forty-nine hand-built pilot cars assembled at Norwood in mid-1966 to evaluate the breed before full production commenced. *GM*

1969

CAMARO Z/28

1 Iconic Chevrolet "cowl induction" power bulge hood signified high performance during the muscle car era.

2 Wide grille opening with dual-plane elements (the grille insert is on a different plane than the opening itself) is dramatic and a first-gen signature feature.

3 Character lines that trailed the wheel openings suggested speed.

4 Simulated air intake grilles weren't functional, but enhanced the 1969 Camaro's performance aesthetic – and became another signature cue of the car.

5 Wide, thin taillamps separated by sheet metal emphasized the first-gen Camaro's width, enhancing its muscular, planted stance.

6 Rear fenders were pulled out, giving the car a wider, more muscular flair.

OFF & RUNNING

2

First generation: 1967–1969

Chevrolet unveiled its first Camaro to the press on September 12, 1966, announcing it would be offered with or without a roof in two model lines, one with a base 230-cid Turbo Thrift six, the other with a standard 327-cid Turbo Fire V-8. A Super Sport rendition (see Chapter 3), powered by an exclusive 350 Turbo Fire small-block, also was predictably introduced to serve as a flagship countering Ford's Mustang GT. When Pete Estes opened the curtain, journalists were treated to a topless SS 350, a stacked-to-the-roof model that demonstrated just how sexy Chevrolet's new compact could get.

"Now perhaps you can understand why we couldn't give it the name of a stealthy, vicious animal any more than one of you would have named an excitingly beautiful woman 'Panther,'" he said. "To continue the feminine analogy, there are plenty of curves to Camaro. In fact, its entire aerodynamic body is molded of curved contours that were polished to perfection in a jet-age wind tunnel."

Before allowing a peek at those curves, Estes spoke of the changing trends that had helped force GM's hand. "Over the years, there has been a shift within the small car market from the family small car to the small sports-type car. It is in the expanding sports car portion of [this] market that our sales position has been weak, and that's where Chevrolet's newest line of small cars—Camaro—comes in."

At curtains' close, Estes mentioned hopes of selling upwards of 300,000 Camaros that first year, a feat that would "restore Chevrolet's penetration of the

(opposite)
Lasting only three years, Camaro's first-generation run produced arguably the most beloved of the breed, 1969's Z28. *GM*

1967 Camaro	
Construction	unitized body w/bolt-on subframe in front
Model availability	sport coupe, convertible
Base price	$2,466 (6-cyl. coupe), $2,704 (6-cyl. convertible), $2,572 (V-8 coupe), $2,809 (V-8 convertible)
Wheelbase	108 inches
Curb weight	2,910 (6-cyl. coupe), 3,165 (V-8 coupe), 3,070 (6-cyl. coupe), 3,325 (V-8 convertible)
Suspension	coil springs & stabilizer bar, front; solid axle w/single-leaf springs, rear (right-side radius rod added in back in December 1966 w/12-bolt rear axle & certain powertrain combos)
Steering	manual (24:1) recirculating ball; 17.5:1 power-assist RPO N40), optional
Wheels	14x5 stamped-steel
Tires	7.35x14
Brakes	9.5-inch drums, front & rear; J52 front discs & J50 power-assist, optional
Base engines	140-hp Turbo Thrift 230 inline-six; 210-hp Turbo Fire 327 V-8
Transmission	column-shifted[1] M15 Saginaw 3-speed manual, std (2.85:1 low w/ six-cylinder, 2.54:1 low w/V-8); heavy-duty M13 Borg-Warner 3-speed, Muncie 4-speed & 2-speed Powerglide automatic, optional
Axle ratio	3.08:1, std w/manual trans; 2.73:1 w/automatic

[1]M11 floor shift, optional (floor shifter also included w/D55 console)

small-car field." Considering that Ford has moved 559,451 1964–1965 Mustangs and another 607,568 1966s, it was clear that he and his cohorts weren't foolish enough to dream unreasonable dreams. The final tally for Camaro's initial model run (including export production) was 220,906, compared to 472,121 for its entrenched rival. Were GM people disappointed? Perhaps a bit, but this opening round still represented a great start for a legacy that surely had nowhere to go but up.

"We don't expect Camaro to outsell Mustang immediately," continued Estes, "but eventually it will. We have plenty of respect for Mustang as an established competitor [that] expanded a market pioneered by Corvette and Corvair. However, we do expect Camaro to cut substantially into Mustang sales in the 1967 model year."

While Mustang offered three body styles (coupe, fastback, and convertible) from 1964 to 1973, 1967–1969 Camaro customers were treated to two, a sport coupe and ragtop. *GM*

Once again like Ford's Falcon-based pony car, 1967's Camaro was, in base six-cylinder form, an affordable, practical everyday driver with just enough sporty appeal to bring young buyers running. Equally familiar was the long list of extra-cost amenities that, also in Mustang fashion, could transform a low-priced, polite F-car into a high-dollar, high-profile hauler in a hurry. "Camaro offers the greatest choice of engines, trims, options and accessories Chevrolet has ever introduced with a new line of cars," added Estes. "The buyer can literally tailor his own automobile"—while also retailoring his/her back pocket to fit a thinner wallet.

At $2,466, a basic 1967 six-packer Camaro cost a mere five extra bucks compared to a '67 six-cylinder Mustang coupe. And for those few dollars more, a Chevy customer got a fistful (literally, the difference wasn't much) of extra room inside, plus slightly bigger brakes and larger standard engines. Mustang's starter motors were a 200-cid six and 289-cid V-8.

1967 Base Turbo Fire V-8	
Type	OHV Gen-I small-block
Displacement	327 cubic inches
Bore	4.00 inches
Stroke	3.25 inches
Horsepower	210 at 4,600 rpm
Torque	320 at 2,400 rpm
Compression	8.75:1
Fuel delivery	Rochester 2-barrel carburetor
Air cleaner	13-inch single-snorkel
Ignition	single-breaker, vacuum-advance distributor
Cooling	17.6-inch 4-blade fan (18-inch 5-blade viscous-drive fan included w/optional air conditioning)
Cylinder block	cast-iron w/5 main bearings
Crankshaft	forged-steel w/2.2993-inch main journals &1.902-inch rod journals
Connecting rods	forged-steel
Pistons	cast-aluminum flat-top
Cam	hydraulic w/310 degrees duration on intake, 320 on exhaust
Rocker ratio	1.50:1
Cylinder heads	cast-iron w/ball-stud rockers & 5.38 cu. in. combustion-chamber volume
Valve sizes (inches)	1.72, intake; 1.50, exhaust
Valve lift (inches)	0.390, intake; 0.410, exhaust
Exhaust system	cast-iron rear-dump manifolds w/2.25-inch pipes & single muffler

NOTE: *For 1967's base six-cylinder, see "Stovebolt Saga," page 52*

FROM THE HORSE'S MOUTH III

"The Camaro should not have been a design success, as it was based on an existing architecture and admittedly hurried to market to address the personal coupe revolution occurring with Baby Boomer customers. However, the first-generation [model] delivered a pure, classic proportion that will forever be regarded as one of the best-looking cars of its time. It was very lean and muscular, with comparatively minor embellishments for high-performance models. That was in contrast to some of the brasher competitors during the muscle car era, and it has helped the first-generation Camaro maintain timeless good looks."

GM Global Design vice president Ed Welburn speaks after receiving the Black Engineer of the Year award on February 7, 2015, in Washington D.C. *Tony Powell photo, courtesy GM*

SHIFT CHANGE

Oldsmobile pioneered (in mass-production terms) completely shiftless driving in America with its Hydra-Matic automatic transmission, introduced for 1940. Unlike GM's "semi-automatic" Automatic Safety Transmission (a 1937–1939 Olds and 1938 Buick option), which still incorporated a conventional friction clutch for stopping and starting, "Hydra-Matic Drive" relied on a fluid coupling mated to a hydraulically controlled planetary gearbox to shift effortlessly through four forward speeds with no need to involve your left foot. Cadillac began offering the Hydra-Matic automatic in 1941, Pontiac in 1948. Buick and Chevrolet officials, however, opted for their own torque-converter designs, with the former's Dynaflow debuting in 1948. Two years later, Chevy's Powerglide became the first optional automatic in Detroit's low-priced field. Though darned durable, it never did fit well behind big, powerful V-8s, not like GM's modern, ultra-strong three-speed automatics—Cadillac's Turbo Hydra-matic and Buick's Super Turbine—which both debuted in 1964. While Chevrolet did begin offering the Turbo Hydra-matic 400 in 1965, it was limited initially to the full-size line's lowest-powered 396 Turbo Jet big-block, leaving the Powerglide to soldier on as Chevrolet's main automatic choice. The big-block-only TH 400 wasn't mated to the top-dog L78 396 until 1969, the same year Chevy rolled out its new TH 350, a three-speed automatic intended for six-cylinder and small-block applications. Though they wore similar name tags, the Turbo 400 and Turbo 350 auto boxes had little in common and were manufactured separately within the GM empire. And unlike its venerable two-speed forerunner, the TH 350 could handle most anything thrown at it, leaving the Powerglide this time to roll off into the sunset. It was finally retired in 1973.

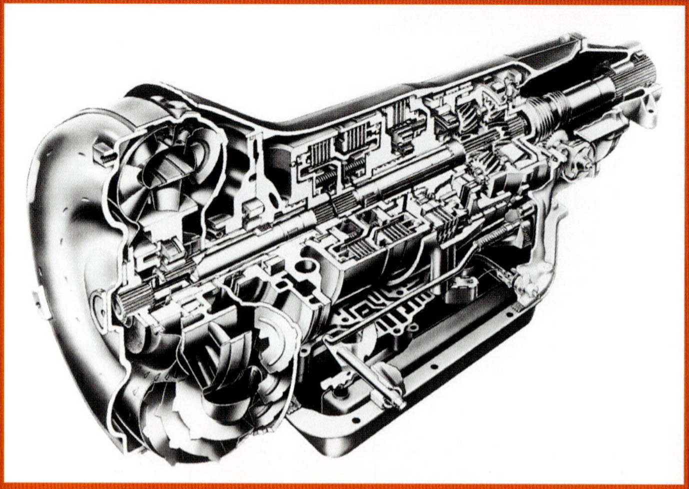

Chevy's Turbo Hydra-matic automatic debuted in 1965 in TH 400 form for the division's Mk IV big-block only. Of a different design, a Turbo Hydra-matic 350 auto appeared for small-block applications in 1969. *GM*

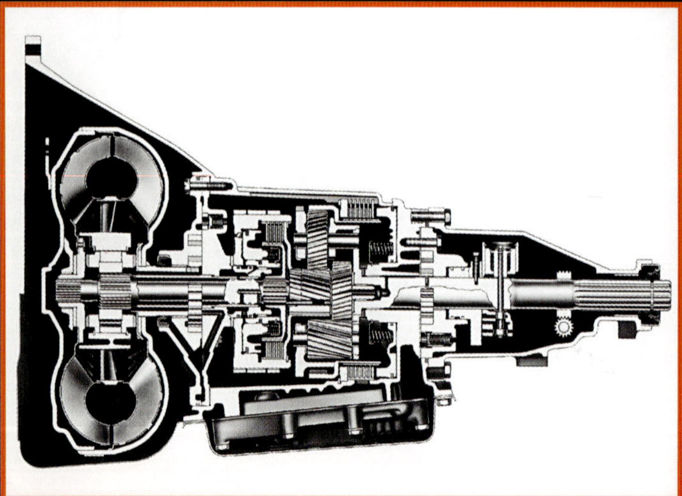

Chevrolet's tried-and-true two-speed Powerglide transmission was the low-priced field's first optional automatic when introduced for 1950. It retired in 1973. *GM*

Also among the original F-car's base package were a column-shifted Saginaw three-speed manual trans and 9.5-inch drum brakes at all four corners. A heavy-duty Borg-Warner three-speed, a Muncie four-speed, and Chevy's yeoman two-speed Powerglide automatic were optional, as were J52 front disc brakes, which came with exclusive Rally wheels—these attractively trimmed rims couldn't be had without those discs, and vice-versa. Power-assist, RPO J50, also was available for (not included with) the J52 discs. Four-piston calipers were the J52 norm in both 1967 and 1968. Still an option for the standard drums in 1968 and 1969, the J50 power booster was included along with the J52 option beginning in 1968.

Also available at extra cost were heavy-duty metallic brakes, a 275-horsepower 327 Turbo Fire (L30), and the Z22 Rally Sport package. That latter deal dressed up both ends of a coupe or convertible with a "distinctive rear tail lamp treatment" and electrically controlled disappearing headlights. RS taillights were solid red with flat-black trim, compared to standard units that used bright trim and incorporated white backup lenses. Rally Sport backup lights were moved down into the rear valance, and a similar move occurred up front for turn signals, which were found in the grille on non-Z22 models.

Missing from this air cleaner lid is the 375-horsepower L78 396 Turbo Jet, a late addition to 1967's Camaro engine lineup. Alert readers might question the 283 Turbo Fire decal seen dead center in this shot. While domestic Camaros weren't fitted with this small-block, Euro-marketed models built in Bienne, Switzerland, were. *GM*

Rally Sports were adorned further with lower body moldings atop black-finished sills (deleted on certain dark-colored cars), bright wheel opening trim and roof drip moldings, and color-keyed upper body stripes. Appropriate badges were included, too, though these were superseded whenever the Z22 option was applied to a Super Sport—"SS" took precedence over "RS."

Buyers who still wanted more needed only wait a few months. Midyear additions included Z28 (see Chapter 4) and two available 396-cid Mk IV big-blocks, the 325-horsepower L35 and 375-horsepower L78, both reserved for Super Sport applications. While the solid-lifter L78 was limited to a stick, the meeker hydraulic-cam L35 could be bolted to either a manual or Chevrolet's three-speed Turbo Hydra-matic automatic, introduced in 1965 along with the division's first Mk IV V-8. Another Camaro SS exclusive early on, D91 nose stripes became available for all other models (save for Z28) in March 1967.

Also appearing just before the turn of the year was a partial response to inherent gremlins that sprung up whenever that Mono Plate rear suspension was tortured by serious torque loads. *Car and Driver's* test crew probably described the situation best after dropping the hammer on a four-speed SS 350 coupe: "Drag racing starts are impossible. The rear axle judders almost

continued on page 54

STOVEBOLT SAGA

Billy Durant's Chevrolet Motor Company originally relied on a T-head six-cylinder to get things rolling in 1911 then turned to a highly successful overhead-valve four-cylinder in 1913. An L-head six appeared in 1915, as did an OHV V-8 in 1917, but neither of these engines proved viable, leaving the pushrod four to lift Chevy to the top of Detroit's sales strata. Bowtie first bested the Blue Oval in 1927, inspiring instant escalations in the battle that still rages today.

Ford's Model A milestone arrived in 1928 and temporarily put the Dearborn gang back on top the following year. Chevrolet then retorted with what *Automobile Trade Journal* called "an epoch in motor car development." The object of this acclaim was the division's new 194-cid six-cylinder, a smooth, efficient OHV powerplant that picked up where Chevy's proven four-banger left off. Although initial prototypes built prior to February 1927 had all been yeoman L-heads, a more advanced OHV six represented the logical step up. After all, the company's familiar slogan had always been "valves in head, ahead in value." New along with this thrifty engine was another catchphrase: "six in the price range of a four."

Nicknames also quickly surfaced. "Stovebolt" became popular after wags noticed the engine's ¼-inch slot-head bolts, cheap-looking choices that helped keep the 1929 Chevy Six's price in line. Cost-conscious iron also was used to cast the pistons, leading to another alias: "Cast Iron Wonder." Both monikers constituted slaps at Chevrolet's tightwad approach, leaving company officials not amused in the least. But the tags stuck, and Stovebolt later even became an affectionate reference that stood for the engine's simple durable nature.

Improvements down the road included an output jump, to fifty horsepower, for 1930. In 1932, that figure rose again to 60 horsepower, the cylinder block was recast for extra strength, and pressure lubrication was instituted for the cam and crank's main bearings. A marked increase to 80 horsepower for the newly named "Blue Streak" six then guaranteed eighty mph for 1934's Chevy. Reportedly, engineers produced those extra horses by tweaking the

All Corvettes built for 1953 and '54 relied on Chevrolet's Blue Flame six-cylinder, which featured split exhausts and three one-barrel carburetors. Blue Flame output was 150 horsepower in 1953, 155 in 1954. *GM*

combustion chambers, and this surely explains why the name was changed to "Blue Flame" in 1935.

A major redesign was performed for 1937 with the idea being to increase durability along with power potential. This second-generation six's cylinder block was recast with extra length, while deck height was shortened by two inches. The latter change went hand-in-hand with a rev-conscious shortened stroke, while the stretched block made extra room for both bigger bores and a switch from three main bearings to four. Displacement increased to 216.5 cid, output to eighty-five horsepower.

Displacement rose again in 1950 after Chevrolet introduced the low-priced field's first automatic transmission, Powerglide. Like most early automatics, Powerglide needed extra muscle to operate as designed, so Chevy officials borrowed the big 235-cid six, used in heavy trucks since 1947, for this automotive application. Advertised at 105 horsepower, the 235 was the first engine in low-priced passenger-car ranks with hydraulic lifters. Manual-transmission cars continued using the solid-lifter, 216.5-cid six.

This same dual-purpose lineup carried over up through 1952, the year Ford unveiled a new 215-cid six-cylinder that *Motor Trend* called "the engine that can and will go places in the years ahead." In comparison, another critic at the time claimed the old Stovebolt was "as up-to-date as red flannel underwear."

Concluding that those undies needed changing, Chevy engineers modernized the Powerglide six for 1953, most importantly by graduating it to full-pressure lubrication. Those cast-iron pistons also were exchanged for lightweight aluminum slugs, effectively ending the Cast Iron Wonder era. It was this engine, tweaked to 150 horsepower, that went into Chevrolet's original Corvette.

Superseded by Chevy's Turbo Fire V-8 in 1955, the company's venerable Stovebolt soldiered on into the 1960s before being upstaged again by a thoroughly modernized third-gen six that featured seven main bearings and shared various V-8 parts and designs, including ball-stud rocker arms. The new Chevy II was the first to put this 194-cid Turbo Thrift six to the test in 1962. Both 230- and 250-cid versions were available for the first Camaro five years later, and that latter inline-six last served in F-car ranks as 1979's base engine. Standard V-6 pony car power then emerged in 1980.

1967 Camaro Six-Cylinder Engines

	Turbo Thrift 230	Turbo Thrift 250
Type	OHV L6 (std.)	OHV L6 (RPO L22)
Displacement	230-cid	250-cid
Bore	3.875 inches	3.875 inches
Stroke	3.25 inches	3.53 inches
Horsepower	140 at 4,400 rpm	155 at 4,200 rpm
Torque	220 at 1,600 rpm	235 at 1,600 rpm
Compression	8.5:1	8.5:1
Fuel delivery	Rochester single-barrel carburetor	Rochester single-barrel carburetor
Air cleaner	13-inch cylindrical housing w/single-snorkel	13-inch cylindrical housing w/single-snorkel
Ignition	single-breaker, vacuum-advance distributor	single-breaker, vacuum-advance distributor
Electrical	45-amp battery & 9-37 amp Delcotron generator	45-amp battery & 9-37 amp Delcotron generator
Cooling	4-blade fan (5-blade fan, 61-amp generator & heavy-duty radiator included w/optional air conditioning	4-blade fan (5-blade fan, 61-amp generator & heavy-duty radiator included w/optional air conditioning
Cylinder block	cast-iron w/7 main bearings & 4.4-inch bore centers	cast-iron w/7 main bearings & 4.4-inch bore centers
Crankshaft	cast-nodular-iron w/2.2993-inch main journals & 2.000-inch rod journals	cast-nodular-iron w/2.2993-inch main journals & 2.000-inch rod journals
Connecting rods	forged-steel	forged-steel
Pistons	cast-aluminum	cast-aluminum
Cam	hydraulic, w/336-degree duration, intake & exhaust	hydraulic, w/336-degree duration, intake & exhaust
Rocker ratio	1.75:1	1.75:1
Cylinder heads	cast-iron w/ball-stud rocker arms & 5.37-cid combustion chamber volume	cast-iron w/ball-stud rocker arms & 5.72-cid combustion chamber volume
Valve sizes	1.715-inch, intake; 1.495-inch exhaust	1.715-inch, intake; 1.495-inch exhaust
Valve lift	0.388-inch, intake & exhaust	0.388-inch, intake & exhaust
Exhaust system	4-port cast-iron center-dump manifold, single reverse-flow muffler w/2.0-inch pipe	4-port cast-iron center-dump manifold, single reverse-flow muffler w/2.0-inch pipe

***NOTE:** L6 is inline six-cylinder*

Multi-carb options disappeared after 1966 from all GM lines save for Corvette, with another exception involving the first-gen Camaro's cross-ram intake (see page 94), available only over dealership parts counters. So was this stillborn triple-carb small-block, tagged L70, intended for Chevy's pony car before the axe fell? It was cancelled in April 1966. *GM*

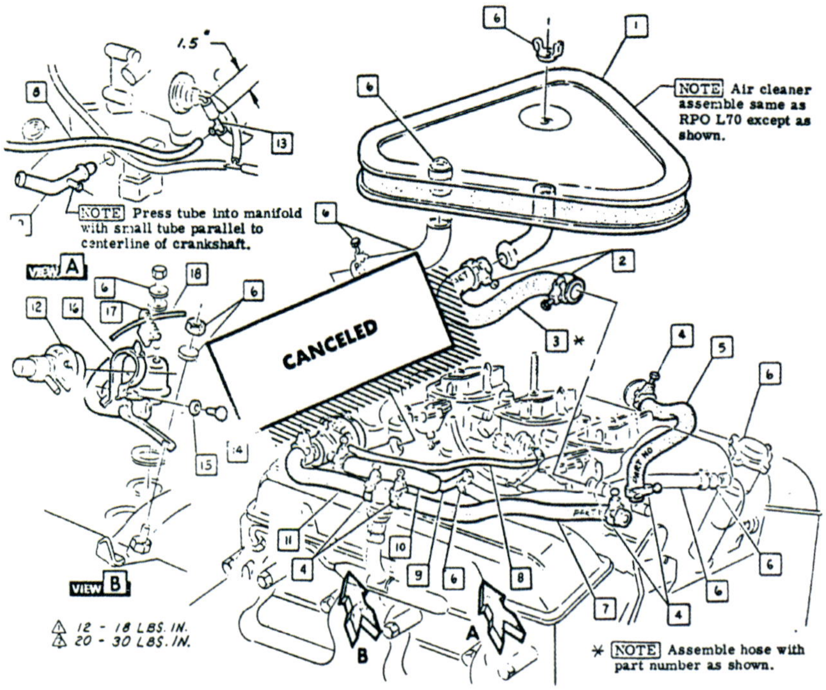

continued from page 51

uncontrollably, with the car hopping sideways almost as far as it is making forward headway."

Simply put, those single-leaf springs couldn't keep the rear axle from "winding up" during hard acceleration, inducing the dreaded high-frequency "wheel hop" phenomenon, which not only impeded headway but was also damned hard on drivelines. The only cure for this malady was to get off the go-pedal ASAP. Not doing so sometimes demonstrated just how easily a U-joint can morph into a hand grenade.

Chevrolet's initial attempt to clamp down on these cagey creatures appeared in December 1966 in the form of a radius rod traction bar attached to the passenger-side end of all heavy-duty twelve-bolt axles, which were installed beneath Z28s and Super Sports, as well as in cahoots with certain drivetrain options. Both round and rectangular rods were tried, with the former appearing first, the latter after February 1967. According to Camaro Research Group (see page 57) investigations, some automatic cars continued receiving round rods after that date, while it's assumed most manual-trans models got the rectangular units.

1967–1969 Production Figures

Year	Body Style	6-cylinder	V-8	Line Total
1967	sport coupe	53,523	142,242	195,765
	convertible	5,285	19,856	25,141
	Total			**202,906**
1968	sport coupe	47,456	167,251	214,707
	convertible	3,513	16,927	20,440
	Total			**235,147**
1969	sport coupe	34,451	190,971	225,422
	convertible	1,707	15,866	17,573
	Total			**242,995**

NOTE: *Totals include both domestic & export production; for Super Sport totals, see Chapter 3; for Z28 totals, see Chapter 4*

While all 1967–1969 Z's received twelve-bolt rears, some Powerglide-equipped SS 350s were sold in 1969 with mundane ten-bolters, which weren't fitted with a right-side radius rod in any application. A twelve-bolt also was added to the L30/four-speed combo beginning February 1967, and most Camaros fitted that year with any engine (six or V-8), optional Positraction, and either the 2.73:1 or 3.55:1 ratio got the H-D axle, too. Additional twelve-bolt installations included Camaros fitted with 1969's rare four-wheel disc brake option (JL8), with that year's new 350-cid LM1 V-8, or with the L65 small-block, another 350 that replaced the LM1 midyear in 1969. All JL8s and manual-trans LM1s, most LM1 automatics, and some L65s featured twelve-bolt rears.

Camaro Engines, 1967–1969

Year	RPO	Displacement (CID)	Horsepower	Torque	CR
1967	std 6-cyl.	230	140 at 4,400 rpm	220 at 1,600 rpm	8.5:1
	L22 6-cyl.	250	155 at 4,200 rpm	235 at 1,600 rpm	8.5:1
	stdV-8	327	210 at 4,600 rpm	320 at 2,400 rpm	8.75:1
	L30 V-8	327	275 at 4,800 rpm	355 at 3,200 rpm	10.0:1
1968	std 6-cyl.	230	140 at 4,400 rpm	220 at 1,600 rpm	8.5:1
	L22 6-cyl.	250	155 at 4,200 rpm	235 at 1,600 rpm	8.5:1
	stdV-8	327	210 at 4,600 rpm	320 at 2,400 rpm	8.75:1
	L30 V-8	327	275 at 4,800 rpm	355 at 3,200 rpm	10.0:1
1969	std 6-cyl.	230	140 at 4,400 rpm	220 at 1,600 rpm	8.5:1
	L22 6-cyl.	250	155 at 4,200 rpm	235 at 1,600 rpm	8.5:1
	std V-8	327	210 at 4,600 rpm	320 at 2,400 rpm	9.0:1
	LM1 V-8[1]	350	255 at 4,800 rpm	365 at 3,200 rpm	9.0:1
	L65 V-8[1]	350	250 at 4,800 rpm	345 at 2,800 rpm	9.0:1

NOTE: *CR is compression ratio; see Chapter 3 for Super Sports V-8s; see Chapter 4 for Z28 V-8s; see Chapter 5 for 1969's COPO V-8s*

[1]LM1 350 Turbo Fire V-8 replaced midyear by L65 350

Though it might've looked good on paper, that radius rod only mildly quieted those gremlins, leading engineers to return to their drawing boards for 1968. They came back with two rear suspension upgrades: multi-leaf springs and

Like Mustang, Chevrolet's new Camaro came standard with bucket seats in 1967. Options pictured here included a four-speed stick, console, AM radio, seat belts, and full instrumentation. Those gauges required both a V-8 and the D55 console. *GM*

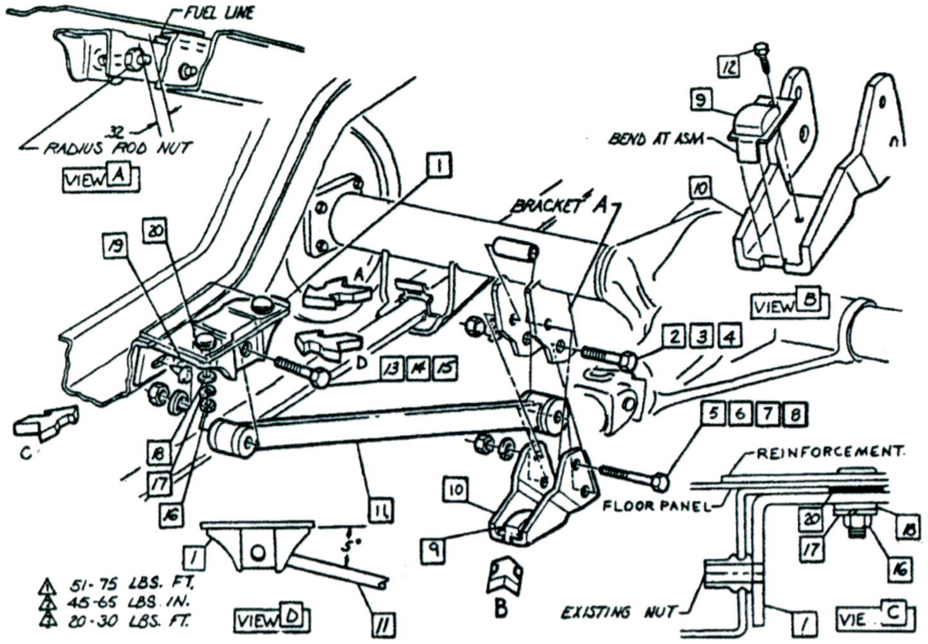

Early Camaros experienced wheel hop problems whenever pedal was mashed to metal, resulting in the midyear 1967 installation of a radius rod to the right end of the rear axle in certain applications. *GM*

staggered shock absorbers. The work of staff engineer Paul King, the staggered shocks (the right located ahead of the axle, the left behind) helped to minimize axle windup, though some journalists continued to complain, especially after driving the highest-performance first-gen Camaro.

"At the mere suggestion of work, the axle leaps and hops, judders and bucks, like the kid who bawls before the switch gets close," announced a *Car Life* report on 1969's L78 SS 396. "Light the tires, and the axle bangs on its mounts like Skinflint John come for the rent. With each fast shift there's the thump of Moby Dick versus whaleboat."

Fortunately, suspension updates in 1968 demonstrated their merits better beneath lesser-powered F-bodies. As it was, *Car Life's* harpooneers seemed to have it in for Chevy's early ponies, slinging arrows seemingly every chance they got. Wonder what they really thought.

Staggered shocks were standard on all Camaros beginning that year, but the new rear springs only went where they were needed most. While all 1968–1969 six-cylinder models continued using 1967's Mono Plates, all

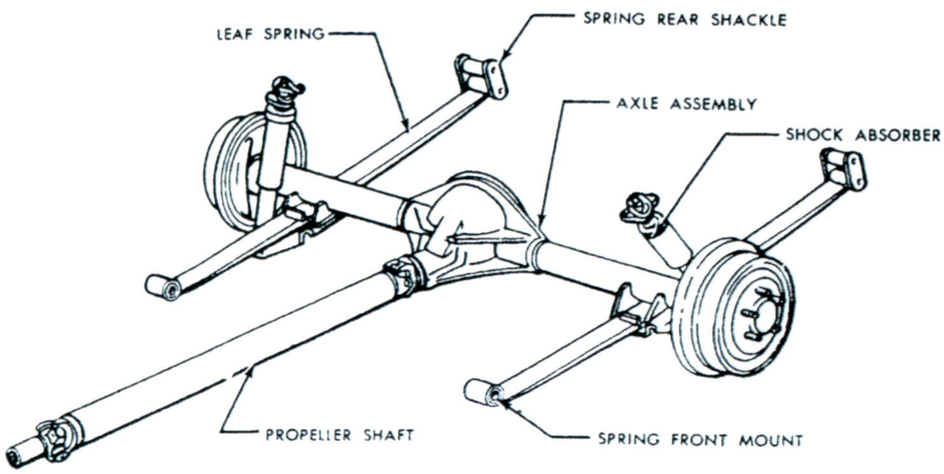

Staggered rear shocks—one mounted in front of the rear axle, the other behind—appeared in 1968 to help cure inherent torque twisting maladies in back. *GM*

JUST THE FACTS, MA'AM

Any gearhead with all his or her bolts in place knows that the internet is literally littered with automotive-oriented websites; many good, but just as many not so much. And, while there certainly are a good deal of fine choices associated with every major model from every brand, Camaro seems to pop up most when the search is narrowed down to quality sites only. Advancing said hunt further, one address stands out well above all the rest: www.camaros.org. Home to the Camaro Research Group, this site concentrates on first-generation models, a complex collection of study subjects if there ever was one. "There have been few other models in the course of automotive history that have been produced in such numbers, with such an incredible, even unbelievable variety of options." explains CRG webmaster Kurt Soren on one of the site's countless informative pages. "Anyone who might mistakenly believe that everything is known about the first-gen Camaro simply hasn't looked into it very deeply!" Look deeply into camaros.org and make sure to take a lunch, or even the next day's breakfast. And if you can't find your answer among the many excellent *CRG Reports* and categorical segments, there's always the group's discussion forum, which seems to bring together only the people who (unlike the bulk of today's blogger types) actually know something about what they're discussing. Along with Soren, you're likely to cross paths with one of the CRG's four co-founders—coordinator Rich Fields, Martin Foltz, Greg Davies, and Wayne Guinn—who put their heads together in September 1997 to get the CRG off the ground. Great job, guys. No, really, *GREAT* job.

twelve-bolters now featured multi-leafs, as did all 350-equipped F-cars (with ten-bolt rears, too) in 1969. Four leafs were used at the tail of all 1968–1969 Z28s (save, apparently for one 1969 model), while five found their way into nearly all other applications. According to *CRG Reports*, it's possible that some Super Sports incorporated four-leaf springs, as did a few LM1, L65, and COPO cars in 1969 (see Chapter 5).

Also new for 1968 was Torque Drive, an option for six-cylinders only. Priced at a paltry $68.65, this intriguing trans (RPO MB1) was simply a Powerglide automatic modified for manual operation. According to promotional paperwork, it "eliminates the clutch [and] offers comparable economy to the regular three-speed manual." MB1 production was 3,099 in 1968, 2,186 in 1969.

"Torque Drive is the greatest thing since the cotton gin," added *Motor Trend's* Eric Dahlquist. "It is easy to work, foolproof, fun and retains some of the physical involvement of manual shifting with almost none of the exertion. It is the familiar two-speed [Powerglide] with the automatic valve body, governor, vacuum modulator, high-speed downshift mechanism and other trifles deleted. The whole deal is very clever, but it begs this

continued on page 61

Made particularly famous by 1969's Indy 500 pace car, Camaro's unforgettable houndstooth interior option debuted in 1968. *courtesy Tom Shaw*

FIRST IN LINE

Overhead-valve V-8s certainly were nothing new when Ed Cole introduced the first of Chevrolet's 100-million-plus small-blocks. Hell, the Bowtie boys themselves originally tried this trick in 1917—and gave up on it in 1918. Their second attempt four decades later, however, resulted in something truly special, a milestone motor still running proud and strong today.

"The introduction of the small-block changed everything," said GM Performance Vehicles and Motorsports vice president Jim Campbell in August 2011. "[It] was an instant success because it offered customers high performance and an affordable price, in a compact design that was very easy to modify. The small-block also started a wave of innovation—and escalating performance—that transformed the cars Americans drove on the street and the track."

With its short (3.0-inch) stroke and innovative ball-stud rocker arms, Chevy's 265 Turbo Fire V-8 could rev with a vengeance. "Here's an engine that can wind up tighter than the E-string on an East Laplander's mandolin, well beyond 6,000 rpm without blowing up like a pigeon egg in a shotgun barrel," wrote legendary *Mechanix Illustrated* scribe Tom McCahill. Instead of rockers all mounted together on a conventional shaft, the Turbo Fire featured lightweight stamped-steel units that each moved independently on its own ball-stud pivot. Along with preserving pounds and energizing rpm, this design meant that deflection wouldn't pass on from one arm to the others via that shaft.

Few seem to remember that this idea actually came from Pontiac's OHV V-8 project, in development dating back to 1946. GM engineer Clayton Leach two years later came up with the ball-stud rocker, which then was implemented in a new Pontiac engine initially targeted for 1953. But various delays pushed that debut back another two years; by then Cole had copped the concept for Chevy's Gen-I small-block.

While maximizing revolutions was a major key to initial success, there are other ways to amplify performance. First, as the old gearhead's adage goes, "there's no substitute for cubic inches." What this means is as plain as the beak on a pigeon's face: in the absence of any other upgrades, increasing displacement represents the quickest, easiest way to increase horsepower.

Gen-I small-block displacement was boosted a third time to 327 cubes in 1962. A 210-horsepower 327 was the base Turbo Fire V-8 for 1967's Camaro. *GM*

Small-block growth began in 1957, and with little fuss or muss. That "440" block (see page 25) ensured there was enough iron present in the block's cylinder walls to allow a bit of boring, an important fact considering Cole expected that rev-intensive stroke to remain as short as possible through whatever growth stages lay ahead. Accordingly, cylinder diameter was punched out from 3.75 inches to 3.875 while the all-important stroke remained at 3.00, resulting in 1957's new 283 Turbo Fire. Next, in 1962, came the 327, which required burning more than a little midnight oil to produce.

According to engineer Maurice Rosenberger, the 283 had barely hit the streets when talk of even more displacement began. A less-than-preferable stroke stretch was a must if more inches were going to happen, but that wasn't the worst of it. Cole wanted a 4.00-inch bore, and that extra cylinder width required serious precision during casting to guarantee suitable cylinder wall thicknesses.

Rosenberger's cohort, Don MacPherson, remembers seeing walls as thin as 0.018 inch, surely the minimum. Early 327s suffered from oil-burning maladies after casting deviations left some walls so lean they deflected during the reciprocating process, allowing lubricants into the combustion area. Casting in a little more iron around the cylinder barrels solved this problem. The 4.00-inch bore then worked fine, and it combined with a 3.25 stroke to equal those 327 cubes.

But now able to make even more power, 327s sometimes destroyed their bottom ends because the carryover crankshaft simply wasn't tough enough. Main bearing journals measured 2.30-inches in diameter, and this was adequately stout as far as the 265 and 283 were concerned, as were the shaft's 2.00-inch connecting rod journals. Chevrolet's bored-and-stroked small-block, however, definitely needed a beefier crank with thickened journals, yet it didn't arrive until 1968, one year before the 327 retired. Measurements for those enlarged main-bearing and rod journals were 2.45 and 2.10 inches, respectively.

Another bottom-end improvement made to the Gen-I involved the crank's main bearing caps. From 1955 to 1967, all small-blocks relied on typical two-bolt caps to hold the crank's five mains firmly in place. In 1968, the L48 began featuring four-bolt caps for the three inner bearings. Additionally, the cylinder blocks for these four-bolt engines were recast with extra iron in their three inner bulkheads to allow all those bolts to torque up tight.

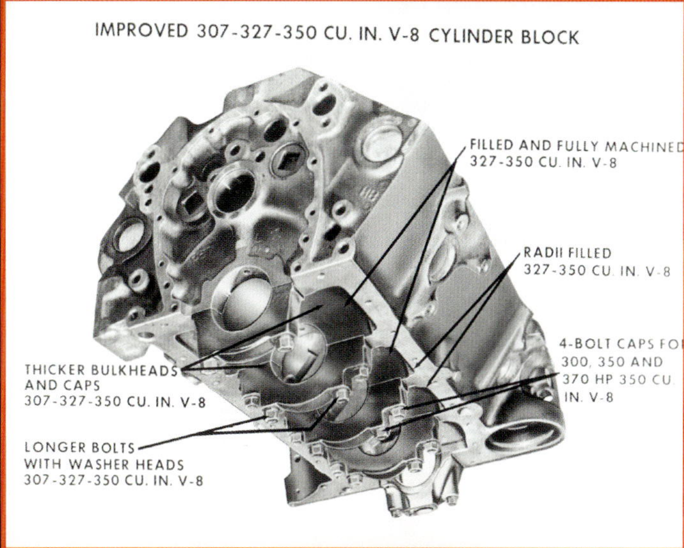

Tougher four-bolt main bearing caps represented just one of the various Gen-I small-block improvements made by 1970. *GM*

The 302-cid solid-lifter small-block was created exclusively for 1967–1969's Z28. Notice the two-bolt main bearing caps near this collection's lower apex. More suitable four-bolt mains appeared in 1969. *GM*

FIRST IN LINE *(continued)*

Both the Z28's 302 (see Chapter 4) and the 350 relied on the 4.00-inch bore. And, introduced as it was all alone for 1967's Camaro SS (see Chapter 3), the 350 also was the first small-block to use the "medium-journal" (an even larger one followed for 1970's 400-cid LF6 V-8) crank. Again, Cole had hoped to avoid any more stroking to create the mightiest mouse to date, but when he told his engineers to achieve that displacement by increasing only cylinder diameters, they had to laugh. As MacPherson later jokingly recalled, the only layout possible within these parameters had the pistons overlapping. There simply wasn't enough space left between the cylinders to meet the boss's demands.

Hence, another new stroke, this one measuring 3.48 inches. And to more confidently handle the inherent extra piston travel, the 350 also was treated to reinforced connecting rods. These rods were predictably heavier, meaning more stress on the crank—without its larger journals, the 350 would have never held together at high rpm. Extra counterweights also were required to balance the new crank, and that added material, working in concert with longer throws, meant that the 350's block had to be modified to allow the bigger shaft ample room to rotate.

Almost lost in the shuffle during the 1960s, the 350-cid Gen I small-block would later rise up during the 1980s to lead Camaro into battle full-force against Ford's Mustang.

Chevrolet's Hottest Small-Block V-8s, 1955–1969

Year	RPO	CID	Bore & Stroke	Horsepower	Torque
1955	n/a	265	3.75 x 3.00	195 at 5,000 rpm	260 at 3,000 rpm
1956	469	265[1]	3.75 x 3.00	225 at 5,200 rpm	270 at 3,600 rpm
1957	579D[4]	283	3.87 x 3.00	283 at 6,200 rpm	290 at 4,400 rpm
1961	354[4]	283[2]	3.87 x 3.00	315 at 6,200 rpm	295 at 5,000 rpm
1962	582[4]	327[2]	4.00 x 3.25	360 at 6,000 rpm	352 at 4,000 rpm
1964	L76	327[2]	4.00 x 3.25	365 at 6,200 rpm	350 at 4,000 rpm
1965	L84[4]	327[2]	4.00 x 3.25	375 at 6,200 rpm	352 at 4,600 rpm
1966	L79	327	4.00 x 3.25	350 at 5,800 rpm	360 at 3,600 rpm
1967	L48	350	4.00 x 3.48	295 at 4,800 rpm	380 at 3,200 rpm
1967	Z28	302[3]	4.00 x 3.00	290 at 5.800 rpm	290 at 4,200 rpm
1969	L46	350[2]	4.00 x 3.48	350 at 5,600 rpm	380 at 3,600 rpm

NOTE: *CID is cubic inch displacement; bore & stroke in inches; CR is compression ratio; engines not identified otherwise all featured single 4-barrel carburetors*

[1]*Dual 4-barrel carburetors*

[2]*Corvette only*

[3]*Camaro Z28 only*

[4]*Fuel injection*

continued from page 57
simple question: Why, since it is very similar to a regular automatic, can this one be marketed for under $70 when the other is a $175 option?"

As for other 1968 updates, exterior changes were predictably minor, with safety-conscious, government-mandated side marker lights added at the corners and 1967's vent windows deleted in favor of new "Astro Ventilation." The popular Rally Sport package returned and once more could be combined with Super Sport gear, but now those hideaway headlights were vacuum-actuated, not electrical.

Little known in 1968 was the aforementioned four-wheel-disc deal, offered initially only over dealers' parts counters (as heavy-duty service parts) beginning in March. Included were 11.75-inch rotors (compared to 11-inch J52 front discs) and four-piston calipers. Even more obscure were a few 1968 test cars fitted with Chevrolet's new Turbo Hydra-matic 350 automatic prior to this trannie's 1969 release as a regular RPO. While both automatics, TH 350 and TH 400, were commonly listed under the same code (M40), the Turbo 350 also answered to M38 when pressed.

Four-wheel discs returned for 1969, this time as a true factory option; RPO JL8, priced at $500.30 for Super Sports and Z28s, $623.50 for other Camaros.

Safety-conscious side marker lights were applied in 1968, helping witnesses pick Chevy's second-edition Camaro out of a crowd, even more so at night. *GM*

Big 15x7 Rally rims were part of 1969's JL8 deal, but only attracted 206 buyers. Most installations understandably went into Z28s, but even some plain-Janers, perhaps even a six-holer or two, may have left the line with four-wheel discs. Like their 1968 forerunners, JL8 discs used four-piston calipers. But 1969's J52 front discs were refitted with simpler, less costly single-piston calipers.

LEADING MAN: JOHN DELOREAN

Best remembered these days as the guy who built the car Michael J. Fox drove back to the future (at 88 miles per hour) in the 1985 movie of the same name, John Zachary DeLorean deserved a much better fate. His gull-winged, stainless-steel DMC-12 sports car, introduced in 1981, might've been a sales flop, but it certainly demonstrated his own penchant for living one step (maybe even two) beyond. "[He] was one of Detroit's larger-than-life figures who secured a noteworthy place in our industry's history," said GM CEO Rick Wagoner a day after DeLorean died on March 19, 2005. "He made a name for himself through his talent, creativity, innovation and daring. At GM, he will always be remembered as the father of the Pontiac GTO." Born in Detroit on January 6, 1925, DeLorean earned a degree in mechanical engineering from nearby Lawrence Institute of Technology in 1948 and then went to work for Chrysler four years later, followed immediately by a stint at Packard, where he rose to research and development director. GM plucked this rising star away from dying Packard in 1956, making him an assistant to Pontiac chief engineer Pete Estes, and he then earned that first notable feather for his cap after helping bring GTO to market in 1964. He was promoted to Pontiac's general manager chair the following year and added a second prominent plumage in 1967 after shepherding Firebird into production. Next came the general manager job at Chevrolet in February 1969, after which time he was asked by the *Detroit News'* Bob Irvin if he could help Camaro outsell Mustang. "I guarantee one thing," he replied, "we're going to try." In October 1972, DeLorean was promoted away from Chevy to direct GM's car and truck group, and many figured he'd eventually rise all the way to corporate president. But he resigned in April 1973 to pursue his own futuristic plans, which involved opening the DeLorean Motor Company in Northern Ireland to build that machine-turned-movie-star. A year before DMC's collapse, DeLorean was busted in Los Angeles in October 1982 while trying to sell $24 million worth of cocaine in a desperate move to fund his failing business. Though acquitted, he never made automotive history again.

Former Chevrolet general manager Pete Estes congratulates former Pontiac general manager John DeLorean after taking over Chevy's top office in February 1969. *GM*

Standard Camaro brakes for 1969 featured new finned drums, and variable-ratio power steering and a body-colored Endura front bumper (RPO VE3) joined this year's options list. As did the Turbo 400 automatic trans—formerly limited to hydraulic-lifter big-blocks—for the cranky 375-horsepower L78 396 Turbo Jet. Both the new Turbo 350 and old Powerglide automatics were available for all of 1969's sixes and small-blocks, and joining that latter lineup this year was a fuel-savoring 350, the aforementioned LM1, rated at 255 horsepower. Effective January 1, 1969, the LM1 was replaced by the 250-horsepower L65 350, and at the same time, the Camaro's standard LF7 327 was retired in favor of Chevy's 307-cid L14 small-block, pegged at 200 horses.

Last, but by no means least, was 1969's cosmetic makeover, which made for a look many still laud today—some even calling it "timeless." Work on this new facade was underway even as Chevy's first pony cars were leaving the gate in 1966, and from the get-go the goal was to drown out some critics' claims that the breed wasn't as macho and muscular-looking as it could be. Or should be.

The Rally Sport package, RPO Z22, was best known for its hideaway headlights. The Z22 price for 1969's Camaro (demonstrated here) was a $131.65. ***Bob McClurg***

A V-shaped egg-grate grille was key up front, while wheel arches were emphasized, in Chevrolet's words, by "trailing windsplits that give an illusion of motion." According to styling studio head Henry Haga, those "brows" were inspired by the fender blisters seen on Mercedes' lovely 300 SL "gull-wing" coupe, a definitely timeless classic introduced in 1954. Enhancing frontal impressions were the Rally Sport's new headlight doors, ribbed transparent panels that the headlights shone through in case anything went wrong with the hideaway process.

Clearly the best Camaro yet, 1969's rendition nonetheless was a one-hit wonder, rolling proof of how well GM people recognized that what makes big news today often is old news tomorrow. Hence, an even better F-car was waiting in the wings while the last of the first generation was turning heads nearly a half century back.

Turn to page 124 to read all about it.

Starting sticker for 1969's V-8 convertible was $2,940. The 210-horsepower 327 that served as the base small-block in 1968 was superseded midyear in 1969 by a 200-horsepower 307. *GM*

Restyled for 1969, RS headlight doors allowed the headlights to shine through should the operating system fail. ***Bob McClurg***

CROWD PLEASER

As part of its 100th birthday celebration in 2011, Chevrolet asked devotees to pick the "Best Chevy of All Time." The winner? The 1969 Camaro attracted 25,058 of 124,368 votes to nose out 1970's Chevelle SS. Also favoring the classic F-car was Chevrolet design director Tom Peters, himself a long-time owner of a 1969 model. "I can vividly remember seeing one for the first time as a kid," he said in September 2011. "The intent of the Camaro was instantly understood, even to a 14-year-old like me, because it possessed a very powerful personality and an elegant, simple design. The Camaro was so cool because it offered this great style, high performance, and yet was attainable for someone just getting out of school." Another 1969 Camaro proud owner, GM global design head Ed Welburn also was full of praise for the car that many feel is not just the all-time top Chevy, but the best automobile ever from any brand. "The 1969 model is the iconic Camaro to me," he added in February 2015. "From the dual-plane grille design and speed lines stamped into the fenders and doors, it was original and distinctive. It didn't borrow from any other design and all these years later, it still looks fresh." According to Peters, it "has become one of the best examples of timeless design in the industry."

Chevrolet Performance Cars design director Tom Peters certainly knows his way around hot-performing Chevys. His pride and joy is a "COPO-inspired" 1969 Camaro powered by an aluminum ZL1 big-block. *GM*

3

SO SHHWEET

Camaro SS: 1967–1972

If there was one major difference between Dearborn's incumbent pony car and Chevrolet's upstart challenger in 1967, it was the various ways in which Pete Estes's package of excitement could be rewrapped. Sure, Ford had a third body style, a fastback, which Chevy didn't. And Mustang GT certainly was no slouch when it came to stirring souls. But if variety truly does spice up life, the new F-car had the advantage, though it may have left more than one interested bystander scratching their noggin' like the flea circus was in town.

"The dilemma facing a potential buyer is a natural one for this time and place in the automotive firmament," explained a March 1967 *Car Life* report. "The problem is not whether to buy the Camaro, but what kind of Camaro, for [this car] probably wears more faces than any other now made."

This claim was tough to deny, especially after Chevy announced the 396 Turbo Jet Mk IV V-8 option in November 1966. Introduced about the same time was the race-ready Z28 (see Chapter 4), and, just like that, it was Ford followers doing all the scratchin' around. In the dust, that is.

Even before the big-block Camaro and Z-car appeared, Chevy fans could still find a decent dose of optional performance, to the delight of dealers everywhere who never met a bill they couldn't run up. In September 1966, the hottest Camaro around was the SS 350, the Super Sport rendition that was available as a sport coupe or convertible. This SS deal, Chevrolet's fourth,

(opposite)
Initially sold only in small-block form, Chevrolet's new Super Sport Camaro became available with the 396-cid Mk IV big-block in November 1966. Two 396 Turbo Jets were offered, the 325-horsepower L35 and 375-horsepower L78, with the latter residing beneath this sport coupe's hood.

Super Sport Family Tree	
1961	Impala SS (6-cyl. & V-8)
1963	Nova SS (6-cyl., small- & big-block V-8s)
1964	Malibu SS (6-cyl. & small-block V-8)
1965	Z16 Malibu (Chevrolet's first SS 396)
1966	Chevelle SS 396
1967	Camaro SS (small- & big-block V-8s)
1967	Impala SS 427
1968	El Camino SS 396
1970	Chevelle SS 454
1970	El Camino SS 454
1970	Monte Carlo SS454
1971	Chevelle SS (small- & big-block V-8)
1971	El Camino SS (small- & big-block V-8)
1983	Monte Carlo SS
1990	454SS pickup
1994	S10 SS compact pickup
1994	Impala SS (Caprice-based)
1996	Camaro SS
2000	Monte Carlo SS
2003	SSR pickup
2003	Silverado SS
2006	Silverado Intimidator SS
2004	Impala SS
2005	Cobalt SS
2006	Malibu & Malibu Maxx SS
2006	Trailblazer SS
2008	HHR SS
2010	Camaro SS
2014	Chevrolet SS

The newest member of the Super Sport clan, Chevrolet SS, debuted for 2014 with a Corvette-sourced LS3 small-block V-8, rated at 415 horsepower. A 2016 model is pictured here. *GM*

included a special hood with simulated vents, red-stripe D70 Wide Oval rubber on wider 14x6JK wheels, and higher rate springs front and rear. Also added to the clean side were various dress-ups, including a color-keyed (black or white) accent stripe around the nose and appropriate identification on the grille, fenders, and fuel-filler cap.

Inspiration for that SS 350 tag came beneath those faux intakes on the hood: Chevrolet's new 350-cid V-8, created exclusively for 1967's Super Sport pony car. Because one couldn't come without the other, the whole package was listed under that small-block's RPO, L48, and identified as the "Camaro SS w/295hp Turbo-Fire 350 cubic-inch engine" option.

Additional codes then appeared after two 396 Turbo Jet big-blocks—the 325-horsepower L35 and 375-horsepower L78—were made available. Prices for the L78 and L35 Super Sports were $500.30 and $263.30, respectively, while the L48 deal added $210.65 to a base V-8 model's bottom line. "SS 350" identification on the grille and gas cap in the L48's case was changed midyear to simply "SS" once the big-block models made the scene.

Ordering a Camaro SS in 1967 was achieved by checking off the RPO code for one of three available V-8s, beginning with the L48 350 Turbo Fire small-block.

THE MYSTERY MOTOR MYSTERY

Casual enthusiasts commonly mix up the Mystery Motor story today, an understandable situation considering automotive journalists back in the 1960s couldn't keep the facts straight, either. That the two competition V-8s Chevrolet developed in 1962 shared displacements also surely didn't help matters, but here's the straight dope: it was the "porcupine-head" 427-cid Mk IIS, not the 427-cube Z11 W-series V-8, that inspired "Mystery Motor" headlines after leaving rivals scratching their heads at Daytona in February 1963. The strip-ready W-engine's tale began on August 1, 1962, when Chevrolet released about twenty sets of aluminum body parts for that year's full-sized flyers. Two weeks later, these lightweight components were joined by another Service Package consisting of 409 engine parts: a hotter cam, radical raised-port cylinder heads, and a two-piece dual-carb intake created to mate up with those taller heads. The lower section of this manifold served as a valley cover, while the high-rise upper half handled fuel delivery chores. All this hot hardware then rolled over into 1963, now as part of the Special Performance Equipment Package, RPO Z11. A stroke increase translated into 427 cubes for the Z11 V-8, which also featured a special cowl-induction air cleaner. Reportedly 57 Z11 Chevys were released for 1963, and more might have made it into racers' hands had GM execs not rained on the parade. The Z11 V-8 was retroactively labeled "Mk I" after the Mystery Motor emerged to wear its Mk II moniker. The Mark-motor lineup read like this:

Created about the same time as the "Mystery Motor," the Mk I racing V-8 was a modified 409 that also was stroked to 427 cubic inches. Tall-port heads and a two-part intake manifold set this W-series big-block further apart from its 409 cousin.

Initially displacing 409 cubic inches when developed in 1962, Chevrolet's Mk II V-8 was immediately stroked to 427 cubes, resulting in a name change: "Mk IIS." NASCAR fans at Daytona in February 1963 knew this engine better by its "Mystery Motor" nickname. *David Kimble cutaway, courtesy GM*

Type	Displacement (CID)	Description	Notes
Mk I	427	409-based "W" V-8	Made hay in super-stock & F/X drag racing
Mk II	409	"Porcupine-head" V-8	Precursor to "Mystery Motor"
Mk IIS	427	Stroked "Mystery Motor"	Turned heads at Daytona in February 1963
Mk III	500	Design study that investigated the feasibility of expanding displacement up to 500 cubes	Never made it off drawing board
Mk IV	396	Available for Corvette (425 hp), Z16 Malibu (375 hp) and full-size Chevys (325 hp) in 1965	Displacement increased to 427 cid in 1966 and 454 cid in 1970

1967 Camaro SS	
Construction	unitized body w/bolt-on subframe in front
Model availability	sport coupe ($2,572 base price), convertible ($2,809 base price)
Price (option only)	$210.65 for L48, $263.30 for L35, $500.30 for L78
Wheelbase	108 inches
Curb weight	3,180 (L48 coupe), 3,435 (L48 convertible), 3,328 (396 Turbo Jet coupe), 3,583 (396 Turbo Jet convertible); H-D 3-speed manual added 22 pounds; 4-speed manual added 7 pounds; Powerglide automatic added 14 pounds; TH 400 auto-trans added 56 pounds
Suspension	heavy-duty; coil springs & stabilizer bar, front; solid axle w/single leaf springs, rear (right-side radius rod added in back in December 1966)
Steering	manual (24:1) recirculating ball; 17.5:1 power-assist (RPO N40), optional
Wheels	14x6
Tires	red-stripe D70 Wide Oval
Brakes	9.5-inch drums, front & rear; J52 front discs & J50 power-assist, optional
Base engine	295-hp L48 350 Turbo Fire V-8 w/SS 350; 325-hp L35 or 375-hp L78 396 Turbo Jet V-8s w/SS 396
Transmission	M15 Saginaw 3-speed manual, std w/350 Turbo Fire V-8; M13 Borg-Warner heavy-duty 3-speed manual, std w/396 Turbo Jet; 4-speed Muncie manual, optional; Powerglide automatic optional w/350 Turbo Fire; Turbo Hydra-matic automatic optional w/L35 396 Turbo Jet (automatic not available w/L78)
Axle ratio	3.31:1 std w/manual-trans L48; 3.07:1 std w/auto-trans L48; 3.07:1 std w/396 Turbo Jet

With a hydraulic cam and 10.25:1 compression, the 295-horsepower L48 was easy enough to live with during civilized, everyday operation was also able to wake the neighbors should push inspire shove. As *Hot Rod's* Eric Dahlquist explained, "The newest extension of the mighty mite 265 will pull [Chevrolet's] fat out of the fire, if indeed it ever fell in." After a little street-wise tweaking, Dahlquist's SS 350 managed an impressive 14.85-second quarter-mile pass topping out at 96.65 miles per hour.

Other tests, however, fell a bit short of *Hot Rod's* results, surely because these cars were run right out of the box, as delivered. *Motor Trend's* Steve Kelley reported a 15.4-second/90-miles per hour quarter-mile for the '67 Super Sport, while *Car Life's* leadfoots were not far behind at 15.8 ticks and 89 miles per hour.

Car and Driver's crew found themselves truly disappointed after requesting a full-dress SS for their thrashing sessions. "The optional 'RS,' or Rally Sport package sounded like it might have something to do with improved roadability, so we asked for that, too. In fact, it consists of the trick grille with concealed headlights and 'distinctive rear lamp treatment.'" Perhaps a little preemptive homework would have helped. Priced at $105.35, the Z22 option was available for either Super Sport, small- or big-block, as were all other Camaro options, including Chevy's new front disc brakes, accompanied by those attractive Rally wheels.

Standard behind the L48 small-block was a three-on-the-tree manual, a Saginaw unit tagged M15. A floor shifter was optional (RPO M11) or included whenever an SS 350 buyer opted for the M13 three-speed, a heavy-duty box supplied by Borg-Warner. Available at extra cost as well were the wide-ratio

When Chevy's 327-cid V-8 was stroked in 1967, the resulting L48 350 Turbo Fire was limited to the Super Sport rendition of the new Camaro. Fed by a four-barrel carb, the L48 produced 295 horsepower. *David Kimble cutaway, courtesy GM*

M20 Muncie four-gear and Chevrolet's venerable two-speed Powerglide automatic. Column shifters also were included in auto-trans applications unless the $47.40 D55 console, complete with its floor-mounted lever, was added into the mix, too.

In big-block ranks, the M13 three-speed was the base trannie behind the L35 and L78, with the M20 also offered behind the former. But in place of the small-block's optional Powerglide was the M40 three-speed TH 400 automatic. No auto was available behind the brutish 375-horsepower 396, nor was the M20; the close-ratio M21 four-speed was the only choice for L78 buyers in 1967.

As expected, checking off L35 or L78 that year transformed a Camaro into a certified Mustang murderer—provided that a Chevy customer knew his/her way around a holeshot. "In quarter-mile trials, the [L35] Camaro showed that it could accelerate very quickly so long as the driver was able to keep his foot far enough off full throttle to prevent time-killing wheelspin after breakaway," claimed *Car Life*. "With the 396 and

continued on page 74

1967 350 Turbo Fire V-8	
Type	OHV Gen-I small-block, RPO L48
Displacement	350 cubic inches
Bore	4.00 inches
Stroke	3.48 inches
Horsepower	295 at 4,800 rpm
Torque	380 at 3,200 rpm
Compression	10.25:1
Fuel	Rochester Quadra-jet 4-barrel carburetor
Air	14.16-inch single-snorkel w/chrome lid
Ignition	single-point, vacuum-advance distributor
Cooling	17.6-inch 4-blade fan (18-inch 5-blade viscous-drive fan included w/optional air conditioning)
Cylinder block	cast-iron w/5 main bearings
Crankshaft	forged-steel w/2.4493-inch main journals & 1.902-inch rod journals
Connecting rods	forged-steel
Pistons	cast-aluminum flat-top
Cam	hydraulic, w/310-degree duration on intake, 320 degrees on exhaust
Rocker ratio	1.50:1
Cylinder heads	cast-iron w/ball-stud rockers & 4.79-cubic-inch combustion-chamber volume
Valve sizes (inches)	1.94 intakes, 1.50 exhausts
Valve lift (inches)	0.390, intake; 0.410, exhaust
Exhaust system	cast-iron rear-dump manifolds w/2.25 pipes & dual mufflers

The SS 396 Camaro was available in three flavors in 1968, as a 350-horsepower L34 Turbo Jet big-block joined the rollover L35 and L78 Mk IV V-8s. An aluminum-head option (RPO L89) also appeared for the solid-lifter L78. A black cove panel in back denoted the Turbo Jet's presence up front.

BIG CUBES

The roots of Chevrolet's Mk IV V-8 trace back to July 1962, when engineer Dick Keinath created a modernized big-block to supersede the fabled 409 in competition circles. Tagged the Mk II, this 409-cid racing mill represented a marked departure from its W-series forerunner, only resembling it slightly on the bottom end while sharing its bore and stroke dimensions. From there all bets were off.

On top was a pair of cylinder heads that breathed like blue blazes. Nicknamed "porcupines," these castings featured canted valves that protruded up from the combustion chambers at varying angles looking very much like the haphazard quills on the back of a pesky . . . well, you know. This innovative arrangement was made possible by GM's proven ball-stud rocker arm design. Not only did this innovation greatly reduce valvetrain weight, it also allowed engineers to place the valves where they would work best. And instead of designing the all-important combustion chamber around a limited, conventional inline valve position, the ball-stud rocker's flexibility gave builders the newfound freedom to shape those chambers almost any way they liked.

Intake valves were located up high near their ports and inclined slightly, making for a straighter flow from intake manifold to combustion chamber. An opposite inclination was applied (to a slightly less degree) on the exhaust end to the same effect. In between, they opened into a "modified wedge" combustion chamber that some classified as a "semi-hemi," and their inclination also allowed for a bit more room in those chambers for larger valves.

These excellent heads helped the Mk IIS ("S" for stroked after the original Mk II was enlarged to 427 cubes late in 1962) make 520 horses, which proved themselves in highly public record-breaking fashion while touring Daytona's superspeedway at nearly 170 miles per hour in February 1963. Too bad the "Mystery Motor" already was doomed by then, a victim of GM's infamous anti-racing edict.

The Mystery Motor was tamed for the street, first by reducing bore to ensure compliance with GM's 400-cid

Chevrolet introduced its new 396-cid Mk IV big-block in three forms early in 1965: the 325-horsepower L35 (shown here), the 375-horsepower L37 (for the new Z16 Malibu), and the 425-horsepower L72, available for Corvette and full-size models. *GM*

maximum for all models except full-sizers and Corvette. Introduced early in 1965, this new big-block (retagged "Mk IV") displaced 396 cubic inches and appeared in three forms: a solid-lifter 425-horsepower brute (L78) for Corvette (plus some full-sizers), the 375-horsepower L37 for the equally new Z16 Malibu in mid-sized ranks, and the comparatively docile 325-horsepower L35 for big Chevys. Chevelle's L37 Turbo Jet was similar to the Corvette 396, save primarily for a less aggressive hydraulic cam in place of the original L78's lumpier mechanical stick. Each featured a beefy bottom end with hefty main bearing webs, large bearing surfaces, and 4-bolt main bearing caps. The crank and connecting rods were of forged steel, the pistons were impact-extruded (forged to you and me) aluminum.

Engineers restored the Mystery Motor's bore for 1966's Corvette and full-size models, treating each to two optional 427 Turbo Jets, the 390-horsepower L36 and 425-hp L72. Super Sport Chevelles customers, meanwhile, could chose from three 396 Turbo Jets, beginning with the base L35, again rated at 325 horsepower. Next up the list was the 360-horse L34, which like the L35 featured hydraulic lifters. Valve sizes and compression also carried over. At the top of the heap was the solid-lifter L78, rated at 375 horsepower for 1966's A-bodies. Compression for this cranky critter was 11:1, and it shared valves with the L72 427.

L78 advertised output remained constant up through 1970, the end of the road for this mechanically cammed big-block bully. And fender badges on Chevelle/Camaro Super Sports continued to read "SS 396" that year despite a minor bore boost performed late in 1969 that pushed the cube count up to 402. Muddying the waters further in 1971 was the "Turbo Jet 400" label stuck on Chevelle engines—a tag that surely confused some customers considering 1970's release of the company's biggest small-block to date, the LF6 400 Turbo Fire V-8.

Introduced as well in 1970 was the supreme Mk IV, the 454 Turbo Jet, available for Corvette and Chevelle but not Camaro. The 402-cid LS3 Turbo Jet was last offered as an option beneath A- and F-body hoods in 1972.

Introduced in Chevelle ranks in 1966, the 375-horsepower L78 featured a 4-bolt block, a lumpy solid-lifter cam, and a big 4-barrel carb on an aluminum intake manifold. Optional L89 aluminum heads also appear here. *Tom Shaw*

1965 396 Turbo Jet Comparison

	L35	L37	L78
Availability	Full-size Chevrolet	Malibu SS 396 (Z16)	Corvette, full-size Chevrolet
Maximum horsepower	325 at 4,800 rpm	375 at 5,600 rpm	425 at 6,400 rpm
Maximum torque	410 at 3,200 rpm	420 at 3,600 rpm	415 at 4,000 rpm
Compression	10.25:1	11:1	11:1
Fuel delivery	Rochester or Holley 4-barrel	Holley 4-barrel	Holley 4-barrel
Air cleaner	single-snorkel	dual-snorkel w/chrome lid	open-element w/black lid
Intake manifold	cast-iron	aluminum	aluminum
Valve covers	painted	chrome-plated	painted
Cylinder heads	cast-iron oval-port	cast-iron rectangular-port	cast-iron rectangular port
Combustion chambers	97cc closed	109cc closed	109cc closed
Pistons	cast aluminum	forged aluminum	forged-aluminum
Main bearing caps	2-bolt	4-bolt	4-bolt
Lifters	hydraulic	hydraulic	solid
Valve sizes (int./exh.)	2.06/1.72 inches	2.19/1.72 inches	2.19/1.72 inches

Super Sport equipment was repackaged under RPO Z27 for 1969. The L48 350 small-block was standard; the L34, L35, L78, and L89 big-blocks, optional. The mag-style wheel covers (RPO N96) seen on this Hugger Orange SS/RS cost $73.75 that year. A "mag-spoke" cover (RPO PA2) also was available. *Tom Shaw*

continued from page 71

torque-converter [TH 400] automatic, it wasn't easy." As for the much meaner L78 rendition, not until Ford released its 428 Cobra Jet V-8, in April 1968, did Mustang stand even the remotest chance against Chevy's hottest Camaro in the stoplight sweepstakes.

Available Super Sport drivetrains rolled over into 1968 nearly unchanged, save for a couple notable additions. First came a third 396 Turbo Jet, the hydraulic-lifter L34, rated at 350 horsepower. Weight-saving L89 aluminum heads also debuted for the mechanical-cam L78, but their heavy price tag ($868.95) inhibited production—only 272 sets were sold. Limited to the L35 big-block only in 1967, the TH 400 also was offered behind 1968's L34, as were the M20 and M21 four-speeds.

1968 396 Turbo Jet V-8s

	L34	L35	L78
Carburetor	Rochester 4-barrel	Rochester 4-barrel	Holley 4-barrel
Intake manifold	cast-iron	cast-iron	aluminum
Cylinder heads	cast-iron oval-port	cast-iron oval-port	cast-iron rectangular port
Combustion chambers	97cc, closed	97cc, closed	109cc, closed
Pistons	cast-aluminum	cast-aluminum	forged-aluminum
Main bearing caps	4-bolt	2-bolt	4-bolt
Lifters	hydraulic	hydraulic	solid
Valve lift (inches)	0.461, intake; 0.480, exhaust	0.398, intake/exhaust	0.520, intake/exhaust
Valve sizes (int./exh.)	2.06/1.72 inches	2.06/1.72 inches	2.19/1.72 inches

The M13 three-speed remained the standard gearbox behind the L35/L34/L78 trio, while the optional M21 stick was exclusive to the latter two. Limited to the L78 alone was the new super-heavy-duty M22 four-speed.

First seen in 1965 as a little-known prototype installation (mostly for Corvette), the M22 was nicknamed "Rock Crusher" in honor of its unmistakable whine, which had to be heard to be appreciated. Inside were beefier gears with heavier teeth cut at a lesser angle compared to the other two Muncie boxes, hence all that noise. As ace transmission builder Paul Cangialosi (www.5speeds.com) points out, hearing an M22 in action is as simple as watching the 1971 cult film, *Two Lane Blacktop*, featuring James Taylor and "Beach Boy" Dennis Wilson. In this nearly forgotten film, the two full-time pop-stars, part-time actors take time out from the *Billboard* charts to campaign a big-block-powered '55 Chevy (the same car, driven two years later by Harrison Ford in *American Graffiti*) cross country in search of drag races. What a great soundtrack.

Production of Indy 500 pace car convertible replicas in 1969 was 3,675, all of them Rally Sport models featuring orange-striped Dover White exteriors. Making this example even rarer is its aluminum-head (L89) 396 Turbo Jet V-8. *Bob McClurg*

GAME CHANGERS

Various influences emerged during the 1960s and 1970s to revise the way American automakers made automobiles, including a newfound national focus on highway safety, initially sharpened thanks to the efforts of Connecticut Senator Abraham Ribicoff. A year's worth of congressional hearings kicked off by Ribicoff in May 1965 resulted in the finger of blame for unforgivable traffic death tolls turning away from so-called reckless American drivers (who'd commonly been targeted up to that point), re-aimed instead at reportedly negligent manufacturers. Those hearings also immortalized East Coast lawyer Ralph Nader, whose legendary literary attack on American car companies (General Motors, most specifically), *Unsafe at Any Speed*, was published in November 1965.

"For over half a century, the automobile has brought death, injury, and the most inestimable sorrow and deprivation to millions of people," began Nader's grim prose. "With Medea-like intensity, this mass trauma began rising sharply four years ago reflecting new and unexpected ravages by the motor vehicle. A 1959 Department of Commerce report projected that 51,000 persons would be killed by automobiles in 1975. That figure will probably be reached in 1965, a decade ahead of schedule."

Nader was present in the gallery on June 24, 1966, when the Senate voted unanimously to pass a bill that would become the National Traffic and Motor Vehicle Safety Act. Also passed unanimously by the House, this landmark legislation was signed into law on September 8, 1966, by President Lyndon Johnson, who proclaimed that "[automotive] safety is no luxury item, no optional extra; it must be a normal cost of doing business."

One month later, another act of Congress established the U.S. Department of Transportation, which began official operation on April 1, 1967. Various other agencies were born of the same legislation to help minimize Nader's "mass trauma," under DOT oversight, until the National Highway Traffic Safety Administration (NHTSA) was created in 1970 by that year's Highway Safety Act.

Among NHTSA responsibilities was creating and enforcing various Federal Motor Vehicle Safety Standards (FMVSS), automotive industry regulations originally enacted in 1967. Adopted in March that year, the first of these, "FMVSS 209," established minimum seat belt requirements. Among other things, body-side marker lights also began showing up on all American cars in 1968 via the FMVSS process. Issued in April 1971, FMVSS 215 specified that 1973 models must feature front bumpers capable of withstanding 5-mph collisions and rear bumpers worthy of 2.5-mph impacts.

Congressional action was again required when it finally came time to build cleaner running engines, a need recognized not long after internal-combustion technology first got cooking. Reportedly the term "smog" was originally heard in London in 1905 in reference to coal smoke pollution. Yet it wasn't until such toxic gases started killing people outright (4,000 in London in 1952, 200 in New York City the following year) that regulatory bodies began considering action to somehow keep the air we all breathe safe and clear.

Washington's Air Pollution Control Act of 1955 basically did little more than provide for research to "consider" the problem. America's first Clean Air Act followed in 1963 to step up the effort to guarantee air-quality standards. In between time, officials in California in 1959 became the first nationally to impose automotive emissions standards, this after Los Angeles basin residents had grown fully fed up with the smog problem there. Federal lawmakers then amended their 1963 federal edict with 1965's Motor Vehicle Air Pollution Control Act, which established this country's first real smog-control deadline. By 1968, all of Detroit's cars would begin curbing emissions or else.

Chevrolet's first major move to control emissions involved its Air Injector Reactor (AIR) system, which pumped ambient atmosphere into the exhaust flow immediately outside each exhaust port to help improve combustion, reducing spent gases in the process. Required on all Chevys sold in California beginning in 1966, the AIR pump was an option (RPO K19) in the remaining forty-nine states for that year and the next. K19 equipment then became standard on all models for 1968, per Washington's latest clean-air standards.

Nineteen-seventy also brought another revised Clean Air Act, as well as the Environmental Protection Agency, which opened for business on December 2. Efforts to further clean up Detroit's act then fully escalated, with additional demands involving the eradication of leaded fuels, both to eliminate toxic by-products and pave the way for the introduction of the next wave of emissions controls: catalytic converter systems.

Leaded fuels fouled "cats," hence the need to switch to unleaded gas. But unleaded and high combustion-chamber squeezes didn't mix, explaining why compression cuts were made across the board in Detroit for nearly all 1971's engines. Horses fell off notably from there and were further shackled after the dreaded catalytics started showing up four years later to inhibit exhaust flow.

Few, however, got the chance to take in a live Crusher concert back in the day. Reportedly only 57 were assembled for Corvette and passenger-car installations in 1965, and the RPO count for 1966 read 15 Sting Rays, 12 L78 SS 396 Chevelles. Priced at $310.70 for 1968's L78 Camaro and Z28, the M22 found 1,277 F-car customers that first year, followed by another 2,117 in 1969, 1,185 in 1970, 1,290 in 1971, and 767 in 1972. Beginning in 1970, the M22 was limited to Z28 applications.

In other 1968 news, the RS package could once again be combined with Super Sport equipment to make for one way-cool Camaro. Exterior SS updates included new SS 396 simulated hood vents, which now contained four equally

Adding weight-saving L89 heads to the L78 396 did not change output. An aluminum intake also was standard atop the 375-horsepower big-block. *Bob McClurg*

Super Sport V-8s, 1967–1972					
Year	**RPO**	**Displacement (CID)**	**Horsepower**	**Torque**	**CR**
1967	L48	350	295 at 4,800 rpm	380 at 3,200 rpm	10.25:1
	L35	396	325 at 4,800 rpm	410 at 3,200 rpm	10.25:1
	L78	396	375 at 5,600 rpm	415 at 3,600 rpm	11.0:1
1968	L48	350	295 at 4,800 rpm	380 at 3,200 rpm	10.25:1
	L34	396	350 at 5,200 rpm	415 at 3,200 rpm	10.25:1
	L35	396	325 at 4,800 rpm	410 at 3,200 rpm	10.25:1
	L78[1]	396	375 at 5,600 rpm	415 at 3,600 rpm	11.0:1
1969	L48	350	300 at 4,800 rpm	380 at 3,200 rpm	10.25:1
	L34	396/402	350 at 5,200 rpm	415 at 3,200 rpm	10.25:1
	L35	396/402	325 at 4,800 rpm	410 at 3,200 rpm	10.25:1
	L78[1]	396/402	375 at 5,600 rpm	415 at 3,600 rpm	11.0:1
1970	L48	350	300 at 4,800 rpm	380 at 2,200 rpm	10.25:1
	L34	402	350 at 5.200 rpm	415 at 3,400 rpm	10.25:1
	L78	402	375 at 5,600 rpm	415 at 3,600	11.0:1
1971	L48	350	270 at 4,800 rpm[2]	360 at 3,200 rpm[2]	8.5:1
	LS3	402	300 at 4,800 rpm[3]	400 at 3,200 rpm[3]	8.5:1
1972[5]	L48[4]	350	200 at 4,400 rpm	300 at 2,800 rpm	8.5:1
	LS3[4]	402	240 at 4,400 rpm	345 at 3,200 rpm	8.5:1

NOTE: *CR is compression ratio*

[1]L89 aluminum heads available for L79 396 Turbo Jet; did not change advertised output

[2]Net-ratings were 210 at 4,400 rpm (horsepower) & 300 at 2,800 rpm (torque

[3]Net-ratings were 260 at 4,400 rpm (horsepower) & 345 at 3,200 rpm (torque)

[4]Net-rated

[5]LS3 V-8 not available in California in 1972

fake carburetor stacks. SS 350 vents were the same seen the previous year on all Super Sport hoods. And in April that year, J50 power-assist became mandatory for SS 396 brakes.

Nineteen-sixty-seven's "bumble-bee" striping up front (listed separately as RPO D91) made an encore appearance, but this time could have been superseded by new stripes that ran over the nose then turned backward down the body sides, ending near the doors' trailing edges. Optional fiberglass front and rear spoilers (D80) also appeared for 1968.

Looking more aggressive all around, 1969's Super Sport attracted additional attention (or not) when topped off up front with the new ZL2 cowl-induction hood, a fully functional unit designed by Larry Shinoda that, in some opinions, worked better than it looked. "[It] doesn't catch the eye the way

Unforgettable orange houndstooth cloth seats were included in the Z11 pace car package in 1969. *Bob McClurg*

Only 600 L78 Super Sport Camaros were built for 1970, the last year for the supreme 396 Turbo Jet. ***Tom Shaw***

forward-facing scoops do, but putting the inlet just ahead of the windshield base means it pulls in more air," explained *Car Life*. "The emissions snag is whipped by a valve that shuts off the cold air until the throttle is 80 percent open." Introduced in mid-December 1968, this $79 lid was limited to SS, Z28, and COPO applications. ZL2 installations for 1969 totaled 10,026.

New, too, were standard power front discs on all Super Sports, a Hurst shifter (for all four-speeds), and a 14x7 "SS" five-spoke sport wheel, borrowed from the SS 396 Chevelle's standard package. The sport wheel was offered early on then cancelled in January 1969 after only 314 sets were sold.

HORSEPLAY

Kicked off during the 1950s, Detroit's horsepower race started slowing in the 1970s, due first to 1971's industry-wide compression cuts. Most power outages seen that year were relatively minimal; output for Chevrolet's L48 small-block, for example, dipped from 1970's 300 horsepower to 270. That same L48, however, greeted customers in 1972 with a 175-horse rating. Where did all those ponies go? Nowhere, actually; that precipitous drop occurred primarily on paper as Chevy promotional people, like all their rivals, ceased advertising "gross" ratings in favor of more honest "net" standards. All those gaudy "brake horsepower" (bhp) ratings seen prior to 1972 resulted from independent engine tests run with no accessories installed, no "smog" equipment in place, nor full exhaust systems bolted up—none of the real-world realities that naturally reduced actual output well before those horses reached the rear wheels. Beginning in 1972, these tests featured "complete" engines running on stands. That is, all belts were in place turning alternator pulleys, air pumps, etc., and spent gases were delivered all the way through stock mufflers. Output ratings were still calculated "at the flywheel," but now all underhood power-robbing variables were entered into the equation, hence the new net aspect. Why Detroit switched from gross to net horsepower ratings was no mystery. Distracting both safety crusaders and insurance agents was one goal. But pulling the wool over customers' eyes surely also played a part. Meeting upcoming emissions standards guaranteed even more power cuts ahead, which hopefully wouldn't appear so drastic once net advertising took root.

Time Slips		
Model	**¼-mile[1]**	**Source**
1967 SS 350[2]	14.85/96.65	*Hot Rod*, January 1966
1967 SS 396 L35	14.5/95	*Motor Trend*, May 1967
1968 SS 396 L35	15.6/92	*Motor Trend*, January 1968
1969 SS 350	15.9/88	*Motor Trend*, March 1969
1969 SS 396 L78	14.77/98.72	*Car Life*, May 1969
1969 SS 396 L89	13.0/108.6	*1969 Supercar Annual*[3]
1970 SS 396 L34	15.3/02.7	*Road Test*, August 1970
1971 SS 396	14.83/96.3	*Hot Rod*, July 1971

[1]*quarter-mile performance in seconds/mph*
[2]*mildly modified test car*
[3]Hi-Performance Cars *magazine*

The Z22 Rally Sport option carried over in similar fashion, with hideaway headlights for one last time, but 1969's SS deal was restructured. Now listed as RPO Z27, it included the familiar L48 backed by a three-speed stick, special suspension, F70 Wide Oval rubber on 14x7 rims, a unique hood, sport striping, and appropriate identification, all priced rather nicely at $295.95. L48 output nosed up to 300 horsepower this year.

The base three-speed, standard in all 1969 SS applications, was a heavy-duty Muncie unit, tagged MC1. Chevrolet's new Turbo Hydra-matic 350 (RPO M38) also became an L48 option (along with the venerable M35 Powerglide) that year, and L78 buyers could opt for the M20 four-speed and TH 400 auto for the first time.

Creating an SS 396 Camaro in 1969 was simply a matter of replacing the Z27 group's 350 small-block with one of the same four optional big-blocks listed in 1968. The aluminum-head L89/L78 combo was again a rarely seen installation, with only 311 released in 1969. Even fewer and farther between was the new JL8 power-assisted four-wheel disc option, available for any model that year.

Chevrolet may not have offered a convertible F-car for 1970, but the Super Sport and Rally Sport (now with fixed headlights) remained on the scene to make sure the second-gen Camaro continued turning heads. Updates to the Z27 package included a black-painted grille and hideaway windshield wipers. Beneath the hood, the base L48 and optional L34/L78 big-blocks carried over, but the L35 didn't. Neither did a three-speed stick or the good ol' Powerglide;

The L48 small-block remained standard for 1970's rebodied Super Sport, while underhood options slipped to two: the L34 and L78 Turbo Jets. *Jim Schild*

it was a Muncie four-speed or Turbo-Hydra-matic auto or nothing at all.

Output for the L48 350 Turbo Fire slipped to 270 horsepower in 1971, but that didn't deter the guys at *Road & Track*, who called Chevy's latest Camaro SS one of the world's ten best cars. Neither the L34 nor solid-lifter L78 returned for this year's big-block Super Sport; in their place was the 300-horsepower LS3 Turbo Jet, which actually displaced 402 cubic inches despite those "SS 396" fender badges. Nearly everything else about 1971's Z27 package repeated from 1970.

1967–1972 Super Sport Production Figures						
	L48	L35	L34	L78	L89[1]	total
1967	29,270	4,003	n/a	1,138	n/a	34,411
1968	12,496	10,773	2,579	4,575	272	30,695
1969	22,339	6,752	2,018	4,889	311	34,932
1970	10,012	n/a	1,864	600	n/a	12,476
1971	6,844	n/a	1,533[2]	n/a	n/a	8,377
1972	5,592	n/a	970[2]	n/a	n/a	6,562

NOTE: Eagle-eyes may notice that 1969's engine count (36,309) does not match that year's total Z27 Super Sport tally. According to various sources (Camaro Research Group, Scott Settlemire, etc.), this discrepancy (1,377 more engines than cars) may well be the result of 1969's COPO build, as these 427-equipped Camaros began life as L78 SS models but were released with no SS identification. Hence, 1969's COPO Camaro count may be hiding within those 4,889 L78 cars.

[1]Aluminum head option for L78

[2]402-cid LS3 big-block V-8

Chevrolet's Super Sport Camaro rolled out for one more year before succumbing to sagging popularity. Power also plummeted, with the base L48 newly net-rated at 200 horsepower, the LS3 Turbo Jet at 240. Adding insult to injury (or was it vice-versa?), California refused to allow LS3 Camaro's entry into the the Golden State because it failed to meet its tough emissions standards. As it was, only 970 of the 6,562 SS Camaros built for 1972 featured the LS3, which was the last big-block installed in an F-car.

In the Super Sport's place in 1973 was the luxury-minded Type LT, detailed in Chapter 6. Yes, the times, they were a-changin'.

Only two engines were available for the last Camaro SS in 1972: the base L48 350 and optional LS3 big-block.

***Road & Track* named Chevrolet's Camaro SS 350 one of the world's ten best cars in 1971.**

4

FROM F TO Z

Z28: 1967–1974

By most accounts, Chevrolet was "out of racing" per corporate-wide decree when Camaro first made the scene. Right. And dinosaurs died about five thousand years ago. True believers in either case needed their eyes checked (former) or checked their brains at the door (latter). Or both. GM execs, meanwhile, simply looked the other way and/or turned deaf ears during the 1960s while apparently insubordinate Chevy engineers continued hammering out hot parts and track-targeted performance packages.

Perhaps hardest to miss was RPO Z28, created to morph 1967's new F-body into a rarin' road racer. In this case, the supposedly taboo target was the Sports Car Club of America's fledgling Trans American Sedan Championship circuit, which opened for business the year before, basically as a Mustang playground. Ford's pony car pioneer ran roughshod over Trans-Am rivals in 1966 and 1967 before the Chevy guys, led by Vince Piggins (see "Quick Thinker," page 86), stepped in to even out the field.

On August 17, 1966, Vince Piggins, then assistant staff engineer in charge of performance product promotion, issued a memo to upper brass outlining his plan to build an SCCA-legal Camaro, a car possessing "performance and handling characteristics superior to either Mustang or Barracuda." To meet

(opposite)
Introduced to the automotive press on November 26, 1966, at Riverside, California, Detroit's first Trans-Am pony car was both a terror at the track and well-rounded boulevard brute. "With the Z-28, Chevy is on the way toward making the gutsy stormer the Camaro should have been in the first place," proclaimed a *Car and Driver* review.

SCCA homologation standards, Piggins's hard-charging pony had to have a back seat (which made it a "sedan"), a wheelbase no longer than 116 inches, and an engine displacing no more than 5 liters (about 305 cubes), and Chevrolet additionally had to offer at least 1,000 examples for public sale. Those technical requirements were a cinch, and the build quota also appeared to be no sweat, at least early on.

"The sales department anticipates a volume of 10,000 such vehicles [for] 1967," wrote Piggins. Such confidence helped make approval a slam dunk, after which Piggins's baby rolled into production wearing only its RPO code as a name tag, a simple label that stuck instead of his original suggestion: "Cheetah."

Quick-thinking Chevy engineers created the Z28's exclusive 302-cid small-block by bolting a 283 crank into a 327 cylinder block. In 1967 and 1968, that block used mundane passenger-car two-bolt main bearing caps on the bottom end. Beefier four-bolt mains were added in 1969. *David Kimble cutaway courtesy GM*

FROM THE HORSE'S MOUTH PART IV

"The 1969 Z28 is a legend, one of the top two or three icons from the performance era. Is it the fastest? No. Is it the rarest? No. But it still has developed its own mystique. A collector has to have a '69 Z28 in his collection. The sublime balance of racing success, unique one-year-only styling, and the driving experience of the high-winding 302, with a glorious exhaust note, has allowed the '69 Z28 to gain followers and fans."

Mecum Auctions consignment director
John Kraman, February 2016

A true car guy, John Kraman has been consignment director at Mecum Auctions since November 2006. He has also done commentator/analyst work for Mecum's television productions since March 2008. *GM*

Staying within that 5-liter limit required looking beyond the Camaro realm, where the smallest small-block offered in 1967 displaced 327 cubic inches. Initial experiments with a high-performance 283 small-block proved adequate, especially in light of the competition's reliance on 289 (Ford) and 273 (Dodge/Plymouth) cubes. But Piggins wanted more, and as long as the rules permitted, he was going to exploit every inch he could get his mitts on. He managed to up the displacement ante to 302 by simply bolting a 283 crank into a 327 block, resulting in a mighty mouse motor that impressed road-testers to no end.

"The very-backdoor word is that ["destroking" the 327] has resulted in a happy and extremely potent screamer," wrote *Sport Car Graphic's* Jerry Titus. *Car and Driver's* critics, meanwhile, simply called it the "most responsive American V8 we've ever tested."

Though its cylinder block was a typical passenger-car unit fitted with two-bolt main bearing caps, the 302's rugged crank was made of forged steel instead of nodular cast iron. Below that crank a windage tray prevented oil sloshing in the pan during hard g's. On top were big-valve heads and an 800-cfm Holley four-barrel carburetor on an aluminum intake. Along with various chrome dress-up pieces, the 302 V-8 also was treated to

Noted Z-code Regular Production Options, 1963–1974

RPO	Description
Z03	Super Sport package for Impala (1963, 1968) and Nova (1963)
Z06	Special Performance Equipment, offered for Corvette only in 1963[1]
Z11	Special 427-cid racing version of 409 V-8, offered 1963 only[2]
Z15	SS 454 package for Chevelle/El Camino, 1970–1971
Z16	Malibu SS 396 model, 1965 only
Z20	SS 454 package for Monte Carlo, 1970–1971
Z22	Rally Sport Package for Camaro, 1967–1974
Z24	Impala SS 427 package, 1967–1969
Z25	SS 396 package for Chevelle/El Camino, 1969–1973[3]
Z26	Super Sport package for Nova, 1969–1976
Z27	Super Sport package for Camaro, 1969–1972
Z28	Special Performance Package for Camaro, introduced in 1967
ZL1	All-aluminum 427 V-8, offered for Corvette in 1969 only[4]
ZL2	Cowl Induction hood for Camaro (1969) and Chevelle (1970–1972)
ZR1	Special Purpose Engine Package for LT-1 Corvette, 1970–1972[5]
ZR2	Special Purpose Engine Package for LS6 454 Corvette, 1971

[1]This code was reused for the LS6-powered Corvette, introduced in 2001.
[2]This drag-racing package (for full-size models) actually appeared late in 1962, but the official Z11 code didn't come into vogue until Chevrolet instituted its RPO system in 1963. Z11 also was used in 1969 for the rare Indy Pace Car Camaro coupe.
[3]Included all engines, small- and big-block, from 1971 to '73.
[4]ZL1 427 V8 also installed in some COPO Camaros in 1969.
[5]This code was reused for the LT5-powered 1990–1995 Corvette and the LS9-powered Corvette, introduced in 2009.

Introduced on Corvettes (all four corners) in 1965, disc brakes appeared in Chevrolet's passenger-car parts bin (for the front wheels only) two years later and brought along soon-to-be-familiar Rally wheels as part of the deal. The 1967 Z28's 15-inch Rally rim differed slightly in offset compared to the same style unit bolted up to Corvettes that year.

QUICK THINKER

Corporate dictates may have barred Chevrolet from the track during most of the twenty-seven years that Vince Piggins worked there, but he apparently never got the memo. The collection of high-perf, competition-conscious models Piggins helped create is notable and includes some of Detroit's greatest muscle machines. "Z" cars were his specialty; along with the Z28, he also was directly responsible for 1965's Z16 Malibu (Chevy's first SS 396 Chevelle) and 1969's ZL1 Camaro, a truly wild pony made possible by the clandestine "COPO" loophole, detailed in Chapter 5.

Piggins was a good friend to Chevy's various "super dealers" back in the 1960s, closely working first with Pennsylvanian Don Yenko to help him roll out his road-racing Stinger Corvair in 1966 using COPO paperwork. In 1968 Piggins put together the COPO #9738 Nova—equipped with a Turbo Hydra-matic-backed L78 396—for Illinois dealer Fred Gibb, who wanted to see Camaro's kissin' cousin kick ass in drag racing's automatic trans classes. Then came the COPO 9561 Camaro (again, see Chapter 5) and COPO 9562 Chevelle in 1969, both introduced initially with high-profile Yenko conversions in mind but also sold in "bone-stock," 427-equipped form through other dealers. The COPO 9560 ZL1 Camaro came about thanks to the efforts of Gibb and drag racer Dick Harrell, another close comrade of the quick thinker who continually kept the backdoor to Chevrolet Engineering open to pro racers intent on waving the Bowtie banner high at the track.

"Piggins was a master at finding the [racing] classes to gain Chevrolet the most promotional value," said Jim Mattison, who worked at the processing end of the COPO pipeline during the 1960s. "But Vince was a low-key, quiet man. If you saw him at a race, you didn't know if he was just another guy nosing around the pits looking for autographs or something. He was very intelligent; he was always talking to the guys, working on new ways to win."

Piggins first learned about winning while directing Hudson's fabulously successful stock-car program during the early 1950s. He then jumped to GM in 1956, where he led the Southern Engineering Development Company (SEDCO), Chevrolet's short-lived "competition wing." After the AMA "racing ban" went into effect in 1957, he was found working under various shingles while quietly developing behind-the-scenes motorsports programs. His home in 1960 was Chevrolet's Engineering Product Information office, followed by the Economy, Safety and Performance section in 1962 and the Product Performance department in 1968. By the time he retired in 1983, his bosses no longer forked tongues (to go along with their deaf ears and blind eyes) when confronted with questions about on-track activities. According to his successor, Herb Fishel, Piggins pioneered the use of "off-road" tags to mask the many purposeful products plainly created contrary to corporate policy during the 1960s and 1970s.

Vince Piggins died in October 1985.

Vince Piggins wore various hats during his twenty-seven-year career at Chevrolet, but the job almost always was the same: help make the Bowtie brand a winner at the track. Here he discusses small-block V-8 merits with NASCAR legend Richard Petty, who traded his familiar Dodge for a Chevy midway through the 1978 season. *GM*

deep-groove pulleys, a heavy-duty radiator, and a five-blade viscous-drive fan. Transistorized ignition was available as a service part, not an RPO.

A rather radical solid-lifter cam helped boost output to 290 horsepower, an advertised number that fooled next to no one. "The 290-hp figure quoted for the Z-28 engine seems ridiculously conservative," continued *Car and Driver's* report. "It feels as least as strong as the 327, 350-hp engine offered in the Corvette." According to Jerry Titus, it was "logical to expect a fully prepared version [of the 302] to produce well in excess of 370 honest ponies." Reportedly the speed merchants at Traco Engineering coaxed more than 500 horsepower from the 302 V-8s they built for Roger Penske's Trans-Am race team.

Along with its exclusive hybrid small-block, RPO Z28 also included a fine supporting cast: Chevrolet's superb F41 suspension, a quick-ratio Saginaw manual steering box, and 3.73:1 rear gears. A Muncie four-speed (either the M20 wide-ratio or M21 close-ratio, both priced at $184.35) was a mandatory option, as were front disc brakes with power assist. Thrown in with those discs were four 15x6 Rally wheels, which along with twin racing

1967 Z28 V-8	
Type	OHV V-8 w/chrome dress-up
Displacement	302 cubic inches
Bore	4.00 inches
Stroke	3.00 inches
Horsepower	290 at 5,800 rpm
Torque	290 at 4,200 rpm
Compression	11:1
Fuel delivery	800-cfm Holley 4-barrel carburetor on aluminum intake
Air cleaner	open-element w/chrome lid (cowl-plenum unit, optional)
Ignition	dual-point distributor (transistorized ignition, optional)
Cooling	five-blade viscous-drive fan, heavy-duty radiator & deep-groove pulleys
Cylinder block	cast-iron w/2-bolt main bearing caps, high-capacity oil pump, windage tray & baffled pan
Crankshaft	forged-steel w/2.299-inch main journals; tuftrided
Connecting rods	forged-steel w/2.0-inch journals & 11/32-inch bolts; shot-peened, heat-treated & magnafluxed
Pistons	forged-aluminum domed w/"half-moon" valve reliefs & press-fitted wrist pins
Cam	mechanical w/346 degrees duration (intake & exhaust) & 118 degrees of overlap
Rocker ratio	1.5:1
Cylinder heads	cast-iron L79 w/pressed-in rocker studs & heavy-duty valve springs; 66cc chamber volume
Valve sizes	2.02-inch intakes, 1.60-inch exhausts
Valve lift	0.485 inch, intake/exhaust
Exhaust system	cast-iron manifolds w/2.25-inch main pipes & low-tone mufflers (tubular headers, optional)

Delivered to dealerships in the car's trunk, the cowl-induction air cleaner seen here tacked about $80 onto the RPO Z28 price in 1967. Tubular headers also were available.

1967 Chevrolet Camaro Z28	
Construction	unitized body/frame
Model availability	sport coupe
Price	Z28 option alone cost $358.10; mandatory options included $184.35 M21 4-speed, $79 J52 front discs & $42.15 J50 power brakes
Wheelbase	108 inches
Weight	3,250 pounds
Suspension	heavy-duty w/coil springs & stabilizer bar, front; solid axle w/leaf springs & right-side trailing link, rear
Steering	manual (24:1) recirculating ball; quicker 20:1 ratio & 17.5:1 power steering, optional
Wheels	15x6 Rally
Tires	7.35x15 Goodyear red-stripe
Brakes	power-assisted 11-inch discs, front; 9.5-inch drums, rear
Engine	302-cid V-8 w/800-cfm Holley 4-barrel carburetor
Transmission	close-ratio M21 4-speed w/Muncie shifter
Axle ratio	3.73:1 (Positraction, optional)

stripes on the hood and rear deck represented the only outward clues to this Trans-Am pony car's true identity in 1967. Soon-to-be-familiar "Z/28" emblems didn't debut until midway through 1968.

Popular options included a fiberglass rear spoiler and the Z22 Rally Sport group. All other available Camaro frills—save for Super Sport treatment and air conditioning—could be ordered along with RPO Z28, but only for a coupe. No convertibles were allowed.

Options exclusive to the Z28 application included tube headers and a fresh-air induction setup. This special air cleaner came boxed in the trunk and featured ductwork that ran from the carburetor to a plenum in the cowl. Ordering RPO Z28 with dealer-installed headers required shelling out an extra $779.40 in 1967. On its own the Z28 option cost $358.10. The price was $437.10 with cowl-induction, and a whopping $858.40 with both fresh-air ducting and headers.

Another rare dealer-installed option also broke corporate rules. Supposedly multiple-carb setups were taboo after 1966 in all GM models save for Corvette,

Z28 buyers initially could have any transmission they liked as long as they favored a Muncie four-speed. All other Camaro options were available inside; witness this dressy Parchment interior with console and gauge cluster.

but leave it to Piggins to promote a cross-ram aluminum intake mounting two Holley four-barrels for the Z28's 302 small-block. The price for this service part was about $500. It was offered up through 1969.

In the end, it was the SCCA's minimum production standard that proved potentially problematic for Chevrolet's first Z28, at least at first glance. Only 602 were built for 1967, inspiring the obvious question: how was it legalized for Trans-Am tracks? Chevy officials played a little numbers game by homologating the more plentiful 350-cid Camaro under Federation Internationale de l'Automobile (FIA, the world governing body over SCCA racing) Group I rules, then qualified that car equipped with RPO Z28 under Group II specifications. Don't understand? Don't feel alone; just nod your head in mock recognition.

Basically the same package rolled over into 1968, with minor updates including new larger center caps for this year's Rally wheels, now shod in Goodyear Wide Tread GT E70 tires. New too was a third transmission choice, the gnarly M22 "Rock Crusher" four-speed, detailed in Chapter 3.

Price changes also made news in 1968 as the RPO Z28 tab increased to $400.25. Adding the plenum air cleaner into the mix bumped that number up to $479.25, but the super rare headers cost the same as they did in 1967. Another dealer-installed option—four-wheel disc brakes—showed up in 1968, this after Trans-Am race teams demonstrated their merits on SCCA road courses. Few of these brake packages were sold over dealership parts counters,

A clandestine creation its first year, Chevy's Trans-Am Camaro was promoted prominently in magazine ads in 1968, helping production surge by more than ten times over. RPO Z28 rolled over with few changes. Rare four-wheel dics brakes were optional for the second-edition model, shown here. ***Jeremy Cliff photo, courtesy Mecum Auctions***

Z28 Production, 1967–1975	
1967	602
1968	7,199
1969	20,302
1970	8,733
1971	4,862
1972	2,575
1973	11,574
1974	13,802
1975	1

Familiar fender badges didn't appear until midyear in 1968. This identification was added to the grille and tail the following year. ***Jeremy Cliff photo, courtesy Mecum Auctions***

and this option then became officially listed on the Camaro RPO list, coded JL8, in 1969.

Chevrolet's nicely restyled Camaro body suited the third-edition Z28 to a T that year, even more so when topped off by the functional ZL2 hood. Nearly all standard features carried over unchanged, and the Rally Sport package was again available.

Notable upgrades included a more durable 302 cylinder block (refitted with four-bolt main bearing caps), a thicker front stabilizer bar, and wider 15x7 Rally wheels, although some early models apparently used 1968's 15x6 rims. Two different tires were installed in 1969; the E70 Goodyear Wide Tread GT was joined by Firestone's Sport Car 200. New too was a more precise, definitely preferred Hurst shifter in place of the clunky Muncie stick used in 1967 and 1968.

Various changes were made to the Z28 equipment group during the year as more than one individual component came and went. Low-restriction chambered exhausts were briefly included in RPO Z28, as was a tachometer, rear deck spoiler, and chrome exhaust tips. Prices ranged from $485.15 to $522.40. One

1970 Chevrolet Camaro Z28	
Construction	unitized body/frame
Model availability	hardtop
Price	$3,794
Wheelbase	108 inches
Weight	3,580 pounds
Suspension	heavy-duty; coil springs & 1.00-inch stabilizer bar, front; solid axle w/leaf springs & 0.625-inch stabilizer bar, rear
Steering	manual (24:1) recirculating ball, std; variable-ratio power-assist, optional
Wheels	15x7 five-spoke
Tires	F60x15 Goodyear Polyglas
Brakes	power-assisted 11-inch discs, front; 9.5-inch drums, rear
Engine	LT1 350-cid Turbo Fire V-8 w/Holley 4-barrel carburetor
Transmission	4-speed manual or Turbo Hydra-matic automatic
Axle ratio	3.73:1 in Positraction differential (4.10:1, optional)

little known variation involved deleting the stripes on the hood and deck lid.

Chevrolet's truly new Camaro body for 1970 enhanced Z28 appeal even further, inspiring *Sports Car Graphic's* Paul Van Valkenburgh to claim "all the parts look and act as though they belong together." And next to the hot-tempered 1969 model, the latest-generation Z was a kinder, gentler machine, a still-strong performer that *Car Life* claimed was "as close to a mild-mannered racing car as the industry has come." The second-gen F-body chassis and better-insulated body meant that, in most cases, that trip was smoother and quieter, while a new power source supplied newfound muscle.

Behind that prow was a 360-horsepower version of the Corvette's new-for-1970 350-cid LT-1 V-8, the greatest small-block yet to scream down the pike. With a much wider, more usable powerband than its 302 predecessor, the 360-horsepower 350 was even more responsive—to both full-throttle bursts and slow-speed operation.

Such relatively civilized compatibility allowed engineers to offer the Turbo Hydra-matic automatic

1970 Camaro Z28 V-8	
Type	OHV V-8 derived from Corvette's optional LT-1 small-block
Displacement	350 cubic inches
Bore	4.00 inches
Stroke	3.48 inches
Horsepower	360 at 6,000 rpm
Torque	380 at 4,000 rpm
Compression	11:1
Fuel delivery	780-cfm Holley 4-barrel carburetor on aluminum intake
Air cleaner	dual-snorkel w/chrome lid
Ignition	single-breaker distributor
Cooling	five-blade viscous-drive fan & heavy-duty radiator
Cylinder block	cast-iron w/4-bolt main bearing caps &windage tray
Crankshaft	forged-steel w/2.45-inch main journals, tuftrided
Connecting rods	forged-steel w/2.10-inch journals
Pistons	forged-aluminum domed TRW w/valve reliefs & press-fitted wrist pins
Cam	mechanical w/317 degrees duration on intake, 346 on exhaust, 96 degrees of overlap
Rocker ratio	1.5:1
Cylinder heads	cast-iron w/screw-in rocker studs, valve spring dampers, pushrod guide plates; 66cc chamber volume
Valve sizes	2.02-inch intakes, 1.60-inch exhausts
Valve lift	0.4586 inch, intake; 0.4850, exhaust
Exhaust system	cast-iron manifolds w.2.25-inch main pipes

A restyled, larger center cap (introduced in 1968) graced the Z28's standard Rally wheels in 1969, and an optional ducktail spoiler appeared out back. Very rare Evening Orchid paint (a 1965 Chevy color) was special ordered for this model. *courtesy Mecum Auctions*

(right)
Included in the ZL2 package was a rubber "doughnut" that sealed the 302 small-block's air cleaner to the cowl-induction hood's underside ducting. Z28 output remained at 302 horsepower in 1969.

(below)
Also new for 1969's Z was an available Special Ducted hood, listed under RPO ZL2. This cowl-induction lid featured appropriate "302" badges when applied to the Z28. Another special-order paint car, this model is the only known 1969 Z28 done in Carolina Blue, a Chevrolet/GMC truck color. ***courtesy Mecum Auctions***

(albeit a beefed-up version with a high-stall torque converter) as a Z28 option for the first time. The M40 TH-400 option cost $290.40 when paired with RPO Z28, while the Muncie four-gear, in either M20 wide-ratio or M21 close-ratio form, cost $205.95.

Chevrolet's 350-cid small-block was allowed entry into the Z28's engine bay thanks to an SCCA rules change. Previously, destroking a production engine to meet the 305-cid Trans-Am limit was not allowed. But this left Dodge and Plymouth out of the picture because no amount of mixing and matching of Chrysler hardware could bridge the gap between the 273- and 318-cid Mopar small-blocks. SCCA officials then changed their minds about shrinking strokes in 1970, a move that allowed the 340-equipped T/A Challenger and AAR 'Cuda to qualify for Trans-Am competition. With that done, it became only

Camaro customers could choose from four unforgettable houndstooth interiors in 1969 featuring black, ivory, yellow, or orange insets. That latter shade was used inside that year's Indy 500 Pace Car replicas.

1967–1974 Z28 V-8s

Year	CID	Bore & Stroke	Horsepower	Torque	Compression
1967	302	4.00 x 3.00	290 at 5,800 rpm	290 at 4,200 rpm	11:1
1968	302	4.00 x 3.00	290 at 5,800 rpm	290 at 4,200 rpm	11:1
1969	302	4.00 x 3.00	290 at 5,800 rpm	290 at 4,200 rpm	11:1
1970	350	4.00 x 3.48	360 at 6,000 rpm	380 at 4,000 rpm	11:1
1971	350	4.00 x 3.48	330 at 5,600 rpm[1]	360 at 4,000 rpm[1]	9:1
1972	350[2]	4.00 x 3.48	255 at 5,600 rpm	280 at 4,000 rpm	9:1
1973	350[2]	4.00 x 3.48	245 at 5,200 rpm	280 at 4,000 rpm	9:1
1974	350[2]	4.00 x 3.48	245 at 5,200 rpm	280 at 4,000 rpm	9:1

NOTE: *CID is cubic inch displacement; bore & stroke in inches*

[1]*Net-ratings were 275 at 5,600 rpm (horsepower) & 300 at 4,000 rpm (torque)*

[2]*Net-rated*

Time Slips		
Model	**¼-mile**[1]	**Source**
67 Z28	14.9 /97	*Car and Driver*, March 1967
68 Z28[2]	13.77/107.4	*Car and Driver*, July 1968
69 Z28	14.34/101.4	*Hot Rod*, January 1969
70 Z28	14.10/99.8	*Popular Hot Rodding*, August 1970
71 Z28	14.40/101.0	*1971 Supercar Annual*[3]
72 Z28	14.5/93.0	*Motor Trend*, October 1971
73 Z28	14.69/96	*Car Craft*, July 1973
74 Z28	15.4/90.54	*Motor Trend*, April 1974

[1]*quarter-mile performance in seconds/mph*
[2]*w/optional dual-carb cross-ram V-8*
[3]Hi-Performance Cars *magazine*

logical at Chevrolet to leave the 302 hybrid behind in favor of the larger 350. In the F-body application, the LT-1 was known on paper simply as the "Z28 V-8."

Like 1969's 302, 1970's 350 featured big-port heads, solid lifters, rugged four-bolt main bearing caps, forged-steel crank, high-volume oil pump, and a baffled oil pan with windage tray. On top, a 780-cfm Holley four-barrel found a home on a revised aluminum intake that allowed more clearance for lower hood lines. A heavy-duty radiator was included, and a beefier 11-inch clutch replaced the 10.34-inch unit previously used for four-speed models.

Brakes were 11-inch front discs, 9.5-inch rear drums, with power-assist made mandatory. Front suspension used the same 1.0-inch anti-roll bar (compared to the standard Camaro's 0.678-inch bar) installed in 1969. Heavier F41 springs and staggered shocks appeared again, but were joined by a new rear sway bar.

A cross-ram intake featuring two Holley four-barrel carbs was offered over dealer parts counters as a service package option for the Z28 from 1967 to 1969. *courtesy Mecum Auctions*

Trans-Am-inspired excitement carried over into the second chapter of the Camaro story. According to *Car Life*, "Despite the added weight and tougher emissions controls, [the '70½ Z28 is] faster than ever, and in a way that makes the car driveable by anybody." Five-spoke sport wheels, still measuring 15 inches across, were new, as was the powerplant behind that pointed prow.

PARTING SHOT

What killed the original Z-car? "Recent noise level tests conducted by Chevrolet engineers indicate that extensive development and engineering work required to meet the more stringent 1975 sound level standards would not be feasible from a cost/benefit standpoint," said Chevy general manager Robert Lund early that year. Initial plans had the next Z28 on hold until January 1975 with hopes of meeting said standards, but it was not to be. "Although we realize a large segment of owners will be disappointed with our decision, the demise of the Z28 resulted from a legislative trend in [certain] areas, which began in 1972. By 1978, these codes will require automobile manufacturers to add sound level equipment to all car lines which could result in a cumulative increase in cost to the U.S. consumer of an estimated $100 per car." That said, a single 1975 Z28 did manage to roll off the Norwood line. How and/or why remains a mystery still. "It was probably a pilot car, but no one knows for sure," said former brand manager and Camaro historian Scott Settlemire in February 2016. "By then, the Z28 was done, because, let's face it, it was a caricature of itself. God only knows what happened to that one car."

Standard for the 1970-½ Z28 was a 360-horsepower variation of the Corvette's 350-cid LT-1small-block. Though still a sold-lifter V-8, this screamer offered a much wider, more "cooperative" (as far as slow-speed operation was concerned) powerband compared to its 302 forerunner, meaning an optional Turbo Hydra-matic auto-trans could be offered to Z28 buyers for the first time. *Tom Shaw*

Priced at $572.95 in 1970, RPO Z28 this time included a rear spoiler, chrome exhaust tips, and restyled 15x7 wheels wearing F60 Wide Oval rubber. Special stripes, "Z/28" identification, a blacked-out grille, and various deluxe trim pieces completed the deal.

In 1971, the Z28 came standard with both front and rear spoilers, and all other important components again rolled over, now wrapped up in a package costing $786.75. A compression cut (from 11:1 to 9:1), however, dropped standard output down to 330 horsepower, signaling the beginning of the end both for the king of the Camaros and muscle cars in general. According to a *Car and Driver* test, quarter-mile time went from 14.2 seconds (for 1970's Z28) to 15.1 clicks for a comparably equipped 1971 model.

Advertised output dropped again slightly to 255 horsepower and spoilers were left out of 1972's standard Z28 package, priced this time at $769.15. But

Chevrolet's original LT-1 V8 was rated at 370 horsepower between fiberglass flanks in 1970. The LT-1 350 remained a Corvette option up through 1972. *GM*

even on the downswing, the Z remained a relative force to be reckoned with. "It's about probably as quick a small-block street machine as you can buy these days short of a five-figure price—except for the LT-1 Corvette," wrote *Motor Trend's* John Christy. "Thanks to its LT-1 engine option, it moves; it produces responsive, go-ahead performance and does it in compliance with the Clean Air Act."

More signs of the times appeared in 1973 as the Z28 V-8's Holley four-barrel and aluminum intake were replaced by the L48 350's Rochester Quadra-jet on a mundane cast-iron manifold. Inside the Z car's heart, civilized hydraulic lifters superseded the solid units used from 1967 to 1972, meaning optional air conditioning could finally enter the equation. Output for this quieter 350 small-block was 245 horsepower.

Now priced at $598.05, RPO Z28 was available this year for the base V-8 Camaro coupe and the new Type LT. When combined with the upscale LT, the Z28 package's price dropped to $502.05 due to various shared features.

Standard horsepower dropped to 330 for 1971's Z28, which looked a lot like 1970's. Repackaging RPO Z28 (making a few previous mandatory options "standard") resulted in a price increase to $786.75.

As in 1970 and 1971, combining the Rally Sport (RPO Z22) and Z28 packages in 1972 traded the one-piece Camaro front bumper for the two small units that did nothing to keep that blacked-out grille out of harm's way. *courtesy Mecum Auctions*

Z28

Along with the base model's mild makeover, rather garish Z28 graphics made news in 1974—the last year for Chevrolet's original Trans-Am pony car. Still rated at 245 horsepower, Chevy's final Z-specific small-block at least was treated to a new, more efficient High Energy Ignition (HEI). All things considered, talk on the street remained favorable for the still-proud Camaro.

Calling 1974's Z28 "a quick, comfortable and fine-handling example of the state of the art in performance cars," *Motor Trend's* John Fuchs had little choice but to look on the bright side as all of Detroit's muscle machines finally rolled off into the sunset. "Its 350 cubic-inch four-barrel V-8, while no contest for the high-performance 302 of earlier Zs, still managed to rip off quarter-mile times in the mid-15s at over 90 mph," he wrote. "Not too shabby for a '74 car."

Fortunately a new dawn awaited Camaro crazies just three years down the road. For that story, turn to Chapter 6.

Camaro customers in 1973 could again combine RPOs Z28 and Z22, and this prestigious performance package in turn could by ordered along with the new LT Camaro, a classy model that superseded the Super Sport rendition that year. *GM*

(opposite)
Prominent graphics were new for 1974's Z, which sold well but still couldn't prevent product planners from putting this legendary pony out to pasture. *GM*

GOODYEAR
U·S·A-1

WHAT RULES?

5

1969 COPO Camaros

In Camaro terms, the optional 396-cid Mk IV V-8 was as big/bad as it got in 1969. While Corvettes and Chevrolet's full-sizers could be ordered direct from the factory that year with the meaner/nastier 427 Turbo Jet, all lesser model lines were limited, per corporate mandate, to engines displacing no more than 400 cubes. F-body buyers who wanted anything more before then were spit out of luck.

Or not.

A small number of 427-powered Camaros were marketed during the 1960s by various "super dealers" (see "Factory Indirect," page 104), with Yenko Chevrolet in Canonsburg, Pennsylvania, making the most noise amidst this select group of pioneering "tuners." A noted Chevy racer with four Sports Car Club of America national titles to his credit, Don Yenko first started dealing in his own personal brand of Bowtie brute force in 1966, making track-ready "Stingers" out of Corvairs with a little assistance from Vince Piggins. One of the one hundred Stingers built that first year by Yenko Sportscars, Inc. (the dealership's performance division) scored its own SCCA class crown with Jerry Thompson driving. Another twenty-eight Stingers were built between 1967 and '69.

In 1967, Yenko Chevrolet also began transforming Camaros into "Super Cars" using the Corvette's L72 427, introduced the year before. Drag racer Dick Harrell built Yenko's first 25 S/C Camaros at his newly opened shop, Dick Harrell's Performance Center, in East St. Louis, Illinois, before production was

(opposite)
Beneath the hood of this Hugger Orange COPO 9561 Camaro is a Corvette-sourced L72 427. Save for the standard ZL2 hood, next to nothing gave away the identity of an L72 Camaro in 1969. Optional spoilers front and rear appear here.

relocated east to Canonsburg. A second-edition Yenko Camaro followed in 1968, this time converted with the aid of the same obscure corporate paperwork used to help build those air-cooled Stingers.

In GM parlance, the process was known as "COPO," short for "Central Office Production Order." Meant primarily for volume customers like trucking firms and police forces, COPOs represented a relatively easy way around corporate red tape while fulfilling special equipment orders. In most cases, a thumbs-up from engineering was all that was needed to transform such requests into reality. Of course, if this process worked so well for power company trucks and constabulary prowlers, it also could be applied to performance automobiles—that is, if Mr. P. had anything to say about it.

Making a COPO happen, however, was by no means a simple task. "There was considerable engineering work put into all COPO requests," said Pontiac Historical Services' Jim Mattison, who was there in the late-1960s when Vince Piggins began using this loophole to build some seriously hot Chevys. "I first

FROM THE HORSE'S MOUTH V

"Ro McGonegal, my staffer at *SS&DI*, actually bought a ZL1 in New Jersey, drove it down to Virginia, and let me drive it once before he left *SS&DI* for *Car Craft*. He drove it across country and sold it when he got to L.A. for more than he paid for it. That car was simply awesome. The ZL1 was introduced after the race Hemi and street Hemi, and while it was a brutal car, it didn't have the brutality of the Hemi cars, in my opinion. Driving a ZL1 on the street was a guaranteed winner against 99 percent of the opposition."

Former* Super Stock & Drag Illustrated *editor
Jim McCraw, February 2016

Jim McCraw was named editor at *Super Stock & Drag Illustrated* magazine in the summer of 1967. He left *SS&DI* late in 1973 and joined the staff at *Hot Rod* soon afterward. *courtesy Jim McCraw*

went to work for Chevrolet in 1966. Then, after graduating college in 1968, I was transferred into the Chevrolet Fleet and Special Services department, where I handled COPOs. I know [Yenko's] 1968 Camaros were [COPO cars] because I did 'em."

Mattison also knew what went on behind the scenes any time a COPO was placed. "Every COPO option had a material sheet that listed every part added and deleted to make the package work," he added in March 2014. "It was very intricate and precise work. I have nothing but great praise for the engineers who worked on our COPO requests. The only 'relatively easy' COPOs were the special paint vehicles that we did for fleet users, as well as individual [orders]."

Digging up COPO roots is not easy; not even Mattison can say when this practice originated. He figures the early Z11 aluminum-fendered drag cars of 1962–1963 must have been COPO-created, and the same perhaps was true for the NASCAR-ready "Black Widow" Chevy of 1957—a distinct possibility considering Piggins had a hand in its development, too.

According to archivist Christo Datini at the GM Heritage Center, no internal records detailing COPO history have surfaced to date. "But I spoke to [former GM engineer and author] Ken Kayser about this, as he has seen a lot of Chevrolet paperwork over the years. He told me a story about one of his research trips to Chevrolet back around 1988–1992 when he was given access to a parts drawing file. He was researching 1956–1957 Corvettes and came across a binder labeled 'COPO.'"

continued on page 106

Various Chevy dealers began swapping Corvette big-blocks into Camaros in 1967, with Pennsylvania-based Yenko Chevrolet gaining the most fame among these pioneering "tuners." Shown here is one of sixty Yenko Camaros built for 1967. *Tom Shaw*

FACTORY INDIRECT

Most Chevy fans recognized the obvious state of affairs between F-body fenders fifty years ago: if the 396 Turbo Jet fit in there, the same was true for its 427-cube big brother. And so it was that, with minimal fuss and/or muss, some owners simply swapped out the former in favor of the latter beneath their nearest proverbial shade tree. Or they opted to let their local dealership handle the dirty work for them—in exchange for more than a few pennies and/or dimes.

Some dealers even branched out themselves, marketing their own private stocks of crossbred Corvette/Camaro. Among the earliest of these forerunning tuners was Dana Chevrolet, located in South Gate, California. In business for barely eighteen months, this hotbed for Bowtie high-performance began offering 427-equipped pony cars hot on the heels of the F-body's introduction. Both the L72 and 435-horsepower L71 were offered as part of the Dana 427 conversion, which brochures described as "the first sensible solution for the enthusiast with a Ferrari taste and a Camaro bankroll."

Credited with first making this swap, however, was Southern California race car builder Bill Thomas (of Cheetah fame), whose high-placed friends back in Detroit included—you guessed it—Vince Piggins. Thomas reportedly had a 427 Camaro prototype up and running even before regular-production versions were finding homes in the fall of 1966.

Based in Southgate, California, Dana Chevrolet was among the first to offer a Corvette-powered Camaro in 1967. Shown here is a 1968 Dana Camaro fitted with a 435-horsepower L71 427.

Chicago's Nickey Chevrolet offered the widest selection of tuner Chevys during the 1960s, swapping all forms of Corvette V-8s into Novas, Chevelles, and Camaros. Along with its L71 427, this 1967 Nickey Camaro also features four-wheel disc brakes.

He then hooked up with Chicago's Nickey Chevrolet to build and market these wild horses.

Founded in 1925 by Edward ("E. J.") and John ("Jack") Stephani, Nickey—known well for its "backward-K" logo—was undoubtedly the largest of the Chevy "super dealers" a half century back, perhaps the biggest factory performance sales and service shop anywhere at any time. By the 1960s, it had grown into a 200,000-square-foot facility that, along with its regular inventory, housed the world's largest collection of "Genuine Chevrolet High Performance Parts."

Nickey's high-perf department opened on Irving Park Road in 1965 and began producing Thomas's 427 Camaro in late 1966, using L88s along with the L72 and triple-carb L71. Chevelle and Nova conversions were offered, too, and in 1970 the 427 was joined by an available 454 Turbo Jet, in either LS5 or 450-horsepower LS6 forms.

Sadly, Nickey Chevrolet encountered various business misfortunes beginning in the late sixties, resulting in the dealership's closure in December 1973. Its speed shop, renamed "Nickey Chicago," relocated soon afterward to Chicago's Milwaukee Avenue but only survived four more years.

Across the big lake east in Grand Rapids, Michigan, was another super-duper dealer, Berger (pronounced "bur-jur")

Dick Harrell marketed his conversions through a distribution network similar to Don Yenko's, and he also swapped Corvette 427s into Chevelles, Novas, and Camaros. This 1969 Camaro sports a Harrell trademark—a Corvette-style stinger hood scoop. *Bob McClurg*

Chevrolet, founded in 1925 by William Berger. Forty-one years later, Nickey Chevrolet's former parts manager, Jim O'Connor, teamed up with William's grandson, Dale Jr., to help propel Berger full-force into the muscle car market. But unlike Nickey, Berger performed few conversions, nearly all by customer request. Selling hot parts was Berger's claim to fame, and to help hawk that hardware the dealership hired twenty-one-year-old Jim Luikens in 1968. Within two years, Luikens was parts manager, and not long afterward Berger stood as Chevrolet's number one performance distributor.

Unfortunately, Luikens left Berger in January 1975 to join Joe Hrudka's Mr. Gasket company. Muscle cars, of course, were all dead by then, so his departure mattered little to a family business that simply shifted gears and kept on keepin' on. Berger Chevrolet remains running strong today, managed by Dale Jr.'s son Matt, still in the same location (2525 28th Street, Grand Rapids, Michigan) it has called home since 1965.

Farther out east on New York's Long Island, Joel Rosen's Motion Performance speed shop teamed up with Baldwin Chevrolet in 1967 to market some of the most outrageous machines seen during Detroit's original tuner era. Among Baldwin-Motion's earliest products was a '67 Camaro powered by an uncrated L88 427. But Rosen didn't discriminate; he was soon shepherding over the "Fantastic Five" once his "Phase III" treatment was applied to Corvettes, Chevelles, full-sizers, and Novas, along with Camaros. Featuring anywhere from 450 to as much as 600 horsepower, a fully loaded Phase III machine almost always could dip well into the 11s at the strip. Setting records in a Baldwin-Motion model was simply a matter of how much a lead-footed customer was willing to spend.

The Fantastic Five morphed into the "Sensational Six" in 1971 after Motion created an outrageous 454 big-block swap for Chevrolet's Vega, something *Motor Trend* editors never possibly could've envisioned when they named GM's new compact their "Car of the Year." These law-bending conversions remained popular up into 1974, when federal tailpipe sniffers finally stepped in, issuing a cease and desist order.

Easily the most aggressive Chevrolet super dealer was New York's Baldwin-Motion firm, a partnership between Long Island's Baldwin Chevrolet and Joel Rosen's Motion Performance speed shop. If it wore a Bowtie, it was fair game for a Motion conversion using any and/or all hot parts available. The 1968 Motion Camaro at left is powered by an L88 427; the '71 at right is a maxed-out "Phase III" 454 model. *Tom Shaw*

continued from page 103

"The binder simply listed numbers in order that they were pulled and assigned as a COPO and the dates they were 'reserved for COPO,'" explained Kayser. "There was not a single word as to what the number was for. It was just so the numbers would not be used by the draftsmen and engineers."

"Ken went on to say that he copied all the numbers related to the 1956–1957 [Corvette]," added Datini in February 2014. "This would indicate that the COPO process was around at least as early as this. Ken believes [COPOs] may have begun sometime after World War II. I inquired with the group that maintains [those parts] drawings today and they do not know of such a binder."

Mysterious origins aside, COPO paperwork proved just the ticket to help make Yenko Chevrolet's transformations a bit easier to perform in 1968. COPO #9737, identified as the "Sports Car Conversion," added heavy-duty cooling, upgraded brakes and suspension, and a 140-mile-per-hour speedometer to a special production run of 396-equipped Camaros, which Yenko then used as bases for its 427 S/C models. As in 1967, out went the stock big-block, in went an L72.

Yenko Chevrolet men made 427 swaps themselves in 1967 and 1968 then turned to Vince Piggins to supply them with factory-direct, COPO-created L72 Camaros in 1969. Rob Clary's 1969 Yenko Camaro (foreground) launches here against the late Peter Simpson in his Ermine White ZL1 at the 2005 Pure Stock Drags. ***Tom Shaw***

Although 1969's COPO Chevelle included some SS treatments (blacked-out grille, twin-bulge hood), it was not a Super Sport. Like its 9561 Camaro cousin, the 9562 Malibu featured an L72 Turbo Jet V-8 beneath that SS hood.

But that was just the beginning. Reportedly, Yenko people began meeting with GM officials in the summer of 1968 to discuss ways to make Super Car production even simpler, the goal being to do away with engine transplants altogether. While ordering 427-equipped Camaros direct from Chevrolet through typical RPO channels was impossible, end-running around corporate roadblocks using the COPO route appeared perfectly doable. Chevrolet general manager Pete Estes—a man who knew a little about circumventing rules after ram-rodding Pontiac's supposedly taboo GTO into production five year before—agreed, giving the go-ahead to Piggins's group to develop yet another special run, this one made up of Corvette-powered 1969 Camaros. As Jim Mattison explained further, a long-running friendship between Don Yenko's father, Frank, and engineering legend Ed Cole (who had risen to GM's presidency in 1967) also surely didn't hurt things when the Pennsylvania gang came knocking in Detroit with a proposition previously deemed out-of-the-question.

The end result was COPO #9561, the code given a factory-built L72 F-car that morphed into a Yenko Camaro S/C with next to no sweat in 1969. Yenko also introduced two other 427 Super Cars that year, one a Chevelle, the second a Nova. While the latter was another swap job, the former was created using a second COPO, #9562. In both COPO cases, the initial idea apparently involved building these L72-equipped F- and A-bodies only for Yenko Sportscar, Inc., but that plan changed once word got out about these clandestine cars—and talk traveled even faster than the machines themselves.

A sponsorship deal with West Coast race car builder Bill Thomas got drag racer Dick Harrell involved in the 427 Camaro conversion business in 1966, leading to a short stay with Nickey Chevrolet in Chicago. After leaving Nickey, Harrell hooked up with Don Yenko to build the first twenty-five 1967 Yenko S/C Camaros.

MAX MK IVs

Chevrolet's Mk IV big-block wasn't even a year old when engineers restored the Mystery Motor's original bore diameter (4.25 inches), transforming the 396 Turbo Jet into the 427 Turbo Jet. Superseding 1965's 425-horse L78 V-8, 1966's 427-cid L72 initially was tabbed at 450 horsepower but ended up also listed at 425 by the time it hit the street. What was up?

According to *Motor Trend*, one elucidation "for this curious state of affairs making the rounds is that, since the congressional safety inquisition [see "Game Changers," page 76], there has been a gentlemen's agreement among parties concerned not to advertise more than 425hp. No official explanations have been offered [for the last-second adjustment], and Chevrolet's sticking to the 425-horsepower figure in its latest literature."

Those who collected that rubbish instantly changed their minds after getting a load of 1966's L72 Sting Ray. "Chevrolet insists that there are only 425 horses in there, and we'll just have to take their words for it," began a *Car and Driver* review of the hottest Corvette yet. "Though we feel compelled to point out that these are 425 horses of a size and strength never before seen by man—horses as tall as houses, with hooves as big as bushel baskets." Backing up that big talk in spades were *Car and Driver*'s road testers, who managed 0–60 in 5.4 seconds, the quarter-mile in 12.8 seconds at 112 miles per hour. "Hoochie-mamma" (with or without the optional exclamation point) simply was not a big enough word.

Offered through typical RPO channels for 1966 only, the L72 was re-rated back up to 450 horsepower (to appease drag racing's rules moguls) the following year when swapped into the first of Yenko Chevrolet's Camaro Super Cars. Back in the regular-production world, big Chevys were limited to the 390-horse L36 from 1968 to 1969, while Sting Ray buyers were treated to various meaner, nastier 427s, including 1967's L71, which used three Holley two-barrel carbs to help make a ceiling-shattering 435 horsepower. Also new that year was the race-ready L88 with its aluminum heads and totally uncivilized makeup. An aluminum-head option (RPO L89) also appeared in 1967 for the L71, but few buyers noticed—only sixteen sets were installed. All three 427 deals—L71, L88, and L89—

Advertised at 450 horsepower, Chevrolet's 454-cid LS6 Mk IV big-block was the highest-rated engine released during Detroit's original muscle car era. It was only available in Super Sport Chevelles in 1970. *David Kimble cutaway, courtesy GM*

Along with the L72 427 Turbo Jet, 1960s tuners also installed aluminum-head L88s and triple-carb L71s in Camaros. Exposed here, the 435-horsepower L71 was a Corvette option from 1967 to 1969. *David Kimble cutaway, courtesy GM*

remained available up through 1969, when they were joined by the exotic all-aluminum ZL1, installed in only two Corvettes that year.

The gloves then came off in 1970, this after Chevy's Mk IV V-8 maxed out at 454 cubic inches and was installed in A-body Super Sports along with Corvettes. Offered for the former, but not the latter, the unapologetic LS6 454 wore the 450-horsepower label deemed politically incorrect four years before. And label makers weren't blowing smoke. In many minds, the LS6 Chevelle was the king of Detroit's original muscle car era.

Too bad Chevrolet officials didn't follow through with purported plans to make the 454 Mk IV a Camaro option.

Chevrolet's Mightiest Big-Block V-8s

Debut Year	RPO	CID	Bore & Stroke	Horsepower	Torque	CR
1965	L78[1]	396	4.094 x 3.76	425 at 6,400 rpm	415 at 4,000 rpm	11:1
1966	L72[1]	427	4.25 x 3.76	425 at 5,600 rpm	460 at 4,000 rpm	11:1
1967	L88[2]	427	4.25 x 3.76	430 at 5,200 rpm	450 at 4,200 rpm	12.5:1
1969	ZL1[3]	427	4.25 x 3.76	430 at 5,200 rpm	450 at 4,400 rpm	12:1
1970	L78[4]	402	4.126 x 3.76	375 at 5,600 rpm	415 at 3,600 rpm	11:1
1970	LS5[2]	454	4.250 x 4.00	390 at 4,800 rpm	500 at 3,400 rpm	10.25:1
1970	LS6[5]	454	4.250 x 4.00	450 at 5.600 rpm	500 at 3,600 rpm	11.25:1
1970	LS6[2]	454	4.250 x 4.00	425 at 5,600 rpm	475 at 4,000 rpm	9:1

NOTE: *CID is cubic inch displacement; bore & stroke in inches; CR is compression ratio*

[1]Corvette & full-size Chevrolets
[2]Corvette only
[3]Corvette & COPO Camaro
[4]Chevelle, Camaro & Nova Super Sports
[5]Chevelle SS only

According to Mattison, he first let the cat out of the bag to Stan Emeritt (of Stan Emeritt Chevrolet in Detroit), who immediately asked the obvious question: why couldn't he take delivery of COPO models, too? Several other fast-thinking, big-time Chevy dealers soon also called with the same query, including Berger Chevrolet in Grand Rapids, Michigan. In the end, anyone with a little insider knowledge could've joined in on the fun. An August 1969 Chevrolet assembly manual addendum helped make that at least partially clear, explaining that "contrary to the implication in the COPO option name shown, [9737] is also applicable to other Chevrolet dealers' orders as well as 'Yenko Sports Cars, Inc.'"

Yenko Chevrolet did the most ordering by far in 1969, with the first 100 COPO 9561 Camaros built going to Canonsburg, followed by nearly 100 more later on, all ordered to be decked out in "Yenko S/C" striping. Reportedly Berger Chevrolet also scored fifty COPO Camaros, plus a lesser number of COPO Chevelles. As for the total 1969 count, that number remains a mystery. What is known is that GM's Tonawanda engine plant turned out 1,015 L72 427s

(193 automatics, 822 four-speeds) coded for F-body installations that year, but how many of those actually went into Camaros sold to the public is undocumented.

Also included in the COPO 9561 "High Performance Unit" package were the functional ZL2 hood, a heavy-duty "curved-neck" Harrison radiator, beefed springs, and a special twelve-bolt axle with 4.10:1 Posi gears. Mandatory extras included J52 power front discs and 14x7 XT steel wheels adorned with mundane "dog-dish" hubcaps. Fifteen-inch Rally rims, shod in E70 Goodyear rubber, were available via 1969's COPO 9373 Sports Car Conversion deal, which added another $184.70 into the equation. Optional 4.56:1 cogs also were offered, and a customer could additionally opt for either a Muncie four speed or heavy-duty Turbo Hydra-matic. Column shifts reportedly were the norm in TH 400CX installations, but apparently not all COPO Camaros were fitted with the curved-neck radiator. Total cost for the COPO 9561 option in 1969 was "only" $489.75, compared to the heftier $627.75 asking price for that year's 375-horse L78 396.

continued on page 114

In 1968, Illinois Chevy dealer Fred Gibb (standing in this rare, blurry photo) noticed that drag racing's super-stock classes were devoid of capable auto-trans Chevrolets, thanks in part to the absence of an automatic option for the new 375-horsepower Nova SS. He then contacted Vince Piggins, who concocted COPO #9738, which added GM's Turbo Hydra-matic 400 into the L78 Nova mix. Gibb Chevrolet raced one of these COPO Novas (also shown here), with Dick Harrell doing the driving. *courtesy Bob McClurg*

YENKO CHEVROLET

Don Yenko's father, Frank, began his sales career dealing Durants from his Central Garage in Bentleyville, Pennsylvania, in 1928. Six years later he switched to Chevrolets and in 1949 opened a second dealership in nearby Canonsburg. Born in 1927, Don didn't compete in his first race, driving a Corvette, until he was thirty but quickly made up for lost time. He won Sports Car Club of America B-Production national championships, also in Corvettes, in 1962 and 1963. And like all successful Chevy racers back then, the younger Yenko of course became acquainted with Vince Piggins, who never failed to leave the Engineering department's backdoor wide open to people who knew how to make Bowtie products fly. Piggins, of course, helped him create those legendary big-block "S/C" models, which were superseded in 1970 by his Yenko Deuce, an explosive small-block Nova that didn't attract the attention of insurance agents the way his 427-equipped, COPO-created A- and F-bodies did in 1969. Plans to continue the Yenko Nova legacy into 1971 faded away, for various reasons, as did the proposed "Super Z" Camaro. And 1971's Turbo-Stinger Vega, while a valiant try, surely showed that the great American muscle car was on its last treads. Yenko Chevrolet's final conversion, its Turbo Z Camaro, came and went in 1981. By then located in McMurray, Pennsylvania, about 5 miles northeast of Canonsburg, the family dealership was sold off in 1982. Five years later, Don Yenko was killed when his Cessna 210 crashed in West Virginia. Frank Yenko died, at age ninety-seven, in April 1997.

Yenko Chevrolet offered one last Camaro conversion, the Turbo Z, in 1981. As the name implied, power came from a turbocharged small-block. *Tom Shaw*

Dick Harrell (left) went from amateur drag racer to professional in 1962, campaigning mostly on American Hot Rod Association tracks. Harrell drove for various Chevrolet "super dealers," including Don Yenko, at right. *GM*

ALUMINUM WONDER

Vince Piggins was right back at it a year after he convinced Chevrolet's brain trust to transform their all-new Camaro into a Trans-Am terror. In 1968, he began lobbying his bosses to fund an aluminum big-block racing engine project, with the intention being to support, among others, Bruce McLaren's successful Canadian-American Challenge Cup team. Small-block Chevy-powered McLaren sports-racers had begun their domination of the unlimited Can-Am series in 1967. Then Jim Hall built an aluminum big-block for his Chaparral race team, inspiring McLaren to threaten to look to Ford for a comparable lightweight big-block to power his 1968 cars. Piggins stepped in, promised Bruce his aluminum rat motor, and the rest is racing history. Armed with the ZL1, McLaren destroyed all Can-Am comers from 1968 to 1971, winning thirty-two of thirty-seven events.

Thoroughly baptized by fire on 1968's Can-Am circuit, a "regular-production" ZL1 made its way the following year onto the Corvette's RPO list, as well as into Camaros via Piggins's COPO pipeline. Although some references identified the ZL1 V-8 as a "special L88," creating the king of the 427s was by no means a simple matter of trading cast iron for additional aluminum. While the ZL1 and L88 wore similar heads in 1969, the weight-saving block was a truly unique piece of engineering.

Winters Foundry was responsible for casting the all-aluminum ZL1's block, heads and intake, while machine work and assembly was completed at GM's Mk IV engine plant in Tonawanda, New York, under a "100-percent parts inspection" policy in production areas that Corvette chief engineer Zora Arkus-Duntov described as being "surgically clean."

That block was cast with thickened walls and beefed-up main webs to compensate for the aluminum's weaker nature. Extra fortification also went into the deck to allow for the use of two extra cylinder head hold-down bolts, and trapped in the bores by the heads were eight cast-iron cylinder sleeves. At the bottom end was a fully nitrided forged-steel crank held in place by four-bolt main bearing caps. The same magnafluxed connecting rods introduced midyear in 1968 for the L88 were used.

Humorously rated at 430 horsepower, 1969's ZL1 427 featured a cylinder block and heads cast from aluminum. Eagle-eyes also might notice the Air Injection Reactor emissions plumbing, yet another joke—A.I.R. equipment surely disappeared as soon as a ZL1 Camaro rolled off the lot. *David Kimble cutaway, courtesy GM*

ZL1 heads were based on the open-chamber units introduced midway through 1968 for the "Second Design" L88. Combustion chambers in these heads were opened up (thus the name) around the spark plug. Results of this change included a drop in compression (from 12.5:1 to 12:1) and 30 percent better breathing. Although the large rectangular intake passages remained the same size as the first-generation L88's, they were recontoured internally to help speed air/fuel into those open chambers.

Exhaust ports, meanwhile, were radically reshaped from rectangles into free-flowing round passages. That these sewer-sized ports didn't match up to the 396 manifolds (with their conventional rectangular inlets) installed at Tonawanda mattered not all considering that ZL1 Camaro customers would surely bolt up aftermarket-supplied tubular headers right after their prized pony rolled off the truck—just what you'd expect for a car intended to go right to the track.

Like the open-chamber L88 head, the ZL1 unit featured 2.19-inch intake valves and enlarged 1.88-inch exhausts. The all-aluminum 427, however, was fitted with new TRW forged-aluminum pistons with extra thick tops and strengthened pin bosses. These beefier slugs not only proved more durable, they also re-instated the L88's original 12.5:1 compression by way of their heightened dome area.

The solid-lifter ZL1 cam was even more radical than the L88's, at least as far as lift was concerned. Tests, however, demonstrated that decreased durations cooperated better with those free-flowing ports. The resulting shorter-duration cam, working in concert with the various reinforced internals, helped the ZL1 wind out like no big-block on this planet. Short bursts to 7,600 rpm were within reason.

This screaming yellow zonker is one of only two ZL1 Corvettes built for 1969. ZL1 Camaro production that year was sixty-nine.

1969 COPO Chevrolets			
	#9560	#9561	#9562
Model	Camaro	Camaro	Chevelle
Price	$4,160[1]	$490[1]	$4,100 (total)
Wheelbase	108 inches	108 inches	112 inches
Suspension	H-D coil/leaf	H-D coil/leaf	H-D coils front/rear
Wheels	14x7	14x7	15x6
Tires	F70x14	F70x14	F70x15
Brakes	power front discs, rear drums	power front discs, rear drums	power front discs, rear drums
Engine	ZL1 427 V-8	L72 427 V-8	L72 427 V-8
Transmission	4-speed manual or TH 400 auto	4-speed manual or TH 400 auto	4-speed manual or TH 400 auto
Axle ratio	4.10:1 Positraction[2]	4.10:1 Positraction[2]	4.10:1 Positraction[2]

[1]*For COPO option alone*
[2]*Special heavy-duty differential in 12-bolt housing*

continued from page 110

More costly by a big-time butt-load was a second 427 Camaro offered for 1969, this one coded COPO 9560. Essentially all 9561 components were repeated in this deal save for one major (to say the least) exception: in place of the cast-iron L72 was Chevy's ZL1 427, easily the most exotic powerplant to find its way onto the street during Detroit's original muscle car era.

For starters, its cylinder block and heads were cast in weight-conscious aluminum, as was its bell housing and cavernous open-plenum intake. Atop that intake was an 850-cfm Holley double-pumper four-barrel, a monster of a carb meant to fan some serious flames. Laughingly advertised at 430 horsepower like the L88, the ZL1 reportedly made 525 horsepower right out of the crate, and 600 were well within reach after only a little port work.

At about 500 pounds, the ZL1 was GM's lightest big-block by far. Tipping the scales further at roughly 3,300 pounds, a ZL1 Camaro weighed in about the same as a small-block Z28, making it a formidable force at the track, just as planned.

Magazine road testers in 1969 wasted little time demonstrating just how formidable. *Cars* staff managed a 13.16-second, 110.21-miles per hour quarter-mile pass in a four-speed ZL1 Camaro right out of the box. Dick Harrell added headers (with full exhausts) and slicks to an automatic ZL1 and produced an 11.85/119.06 run for *Super Stock*. Uncapping those headers resulted in an 11.64/122/15 time slip, and later tests pushed the outside of the envelope into the mid-10s. In 1971 a ZL1 using twin Holley Dominator carbs set an American Hot Rod Association Pro Stock record of 9.63 seconds at 143 miles per hour.

Harrell, incidentally, was a prime mover/shaker behind the COPO 9560 Camaro's creation. As mentioned, his relationship with Yenko was a short one; by the summer of 1967 he'd hooked up with yet another quick-thinking

1969 ZL1 Camaro Production Figures				
	Gibb Chevrolet		Other dealerships	
Paint	TH 400	4-speed	TH 400	4-speed
Hugger Orange	4	6	1	5
Lemans Blue	4	6	1	2
Cortez Silver	4	6	0	2
Fathom Green	4	6	0	2
Dusk Blue	4	6	0	0
Daytona Yellow	0	0	0	2
Dover White	0	0	0	2
Garnet Red	0	0	0	2
Total: 69	**20**	**30**	**2**	**17**

dealer, Fred Gibb, in LaHarpe, Illinois. First, Harrell helped transform *Little Hoss*, a 1967 Z28 that Gibb Chevrolet raced, into a big winner at the strip. After relocating his Performance Center to Kansas City, Missouri, he took delivery of twenty of the fifty 1968 COPO Novas ordered by Gibb and swapped out their L78 396s for 427s, marketing them, along with a batch of 427 Chevelle conversions, through a tidy dealership network modeled after Yenko's much larger program. Still splitting time between his tuner shop (known as the "Home of the Original 427 Camaro") and drag racing, Harrell was killed in a Funny Car racing accident in September 1971.

Like Yenko, Gibb and Harrell also approached Piggins in the summer of 1968 about 427 Camaro builds, this time involving both iron L72s and aluminum ZL1s. Drag racing rules at the time specified that at least fifty examples of a particular production car be built to legalize that vehicle for stock-class competition. No problem. Gibb ordered fifty COPO 9560 Camaros for his LaHarpe lot, and the first two arrived on New Year's Day, 1969. Another nineteen went to dealers across the country who, like Gibb, quickly found out how tough it was to market this high-strung, high-priced Camaro.

continued on page 119

Gibb Chevrolet's 1969 ZL1 Camaro established an American Hot Rod Association Pro Stock record in 1971, smoking the quarter-mile in 9.63 seconds at 143 miles per hour. *Bob McClurg*

MODERN MANUFACTURED MUSCLE

Corvette-powered Camaro conversions made mucho headlines prior to 1970 then faded away, right along with the factory-direct breed's high-performance identity. This aftermarket approach then found new life a quarter century later once Chevy's heartbeat began thumping again.

Among the earliest modern tuners to restart F-car tweaking was Ed Hamburger's Street Legal Performance, which introduced its Firehawk Firebird in 1991. Five years later, SLP kicked off construction of Chevy's reborn Camaro SS and continued these transformations up through the fourth generation's final act in 2002. The shingle later became Specialty Vehicle Engineering after the fifth-gen pony car came into focus. In 2014, the SLP/SVE gang dug deep into Camaro's past, bringing back the Panther name for a collection of modified models that ranged from 560 to 700 horsepower.

Known more for his awesome Corvette conversions, Reeves Callaway turned his attentions to Chevrolet's other sporty model in 1993. Callaway Cars, founded in 1976, rolled out its 404-horsepower C8 concept Camaro seventeen years later, leading to production of its SuperNatural models, labeled the "World's Baddest Camaro" by *Motor Trend* in 1994. In 2016, Connecticut-based Callaway topped itself with its SC610, a supreme sixth-gen Camaro that, as its name implied, offered 610 supercharged horses. "It's simply an amazing combination of great car and shocking power," beamed Reeves Callaway in January that year.

Reeves Callaway understandably considers "tuner" to be a misnomer, preferring the more accurate "specialist constructor" for businesses like his that team up with OEM (original equipment manufacturer) sources to maximize regular-production muscle then market it through authorized factory-direct outlets—complete with original warrantees. In February 2016, Callaway Cars announced the launch of the most powerful street-legal Camaro then available via select Chevrolet dealerships. As you might guess, Callaway's 2016 SC610 Camaro featured a 610-horsepower supercharged V-8. *courtesy Callaway Cars*

Originally known as Street Legal Performance (SLP) Engineering, Specialty Vehicle Engineering (based in Toms River, New Jersey) announced late in 2014 that it would be honoring the Yenko Super Car legacy with its new Yenko/SC 427 Camaro, featuring a 700-horsepower supercharged 7.0L small-block. *courtesy SVE*

And if that wasn't electrifying enough, perhaps a visit to Hennessey Performance outside Houston is in order. John Hennessey, in his own words, "has been making fast cars faster since 1991," and one look at the following video, www.hennesseyperformance.com/vehicles/chevrolet-2/camaro/camaro-zl1/2012-2015-chevrolet-camaro-zl1-hpe750-upgrade/, will demonstrate just how fast. On October 23, 2012, an HPE750 ZL1 Camaro ran 203.9 mph on a closed highway outside of Austin, Texas. Is "yikes" a big enough word? Don't answer. Because in 2016 Hennessey began offering its supercharged HPE1000, which, as you might guess, featured one thousand horsepower, giving all new meaning to "grand times."

Another speed freak, Matt Murphy, opened GMMG, Inc., in Marietta, Georgia, in 1999 with intentions of supplying certain Chevy dealers with super Camaros. His first client was Berger Chevrolet. Offered from 2000 to 2002, the GMMG-built Berger SS was fitted with a 380-hp V-8 and six-speed manual trans.

Murphy also created two other special-edition models exclusively for Berger just before the F-car went into temporary retirement, one honoring *Hot Rod* magazine, the other paying tribute to drag racer Dick Harrell, who helped get Chevrolet's original ZL1 into production in 1969. While the *Hot Rod* magazine edition relied on the same modified 5.7L LS1 used by the Berger SS, the Dick Harrell car featured 7.0 liters of muscled-up LS small-block.

A third GMMG Camaro also appeared in 2002 to bring back memories of the wild horse Harrell and Fred Gibb let loose thirty-three years before. "Out with a bang" was the description *Super Chevy*'s Kevin Lee used for GMMG's "ZL1 Supercar," which was powered by a 400-horsepower LS6. As in 1969, the production count for 2002's ZL1 was sixty-nine, with 28 of those going to Berger Chevrolet. The third built went to Gibb's widow, Helen, while the last was a 427-cube variation honoring Bill "Grumpy" Jenkins, who raced the original ZL1 back in the day.

A familiar name dating back to Detroit's original muscle car era, Berger Chevrolet in Grand Rapids, Michigan, has offered various Camaro conversions since 2000, including 2002's Dick Harrell Edition, shown here. *courtesy Berger Chevrolet*

John Hennessey, in his words, has been "taking excellent products and maximizing their potential" since 1991. In 2016, Houston-based Hennessey Performance was offering some of the most outrageous Camaro conversions out there, including the supercharged HPE1000, a 1,000-horsepower sixth-gen road rocket. *courtesy Hennessey Performance*

Other GMMG conversions included the Intimidator SS, created in 2001 for Dale Earnhardt Chevrolet in Newton, North Carolina, and the Tom Henry SS, built in 2001 and 2002 for Tom Henry Chevrolet in Bakerstown, Pennsylvania. The Georgia firm's last super Camaro, the Performance Edition SS, also was released just before the fourth generation ended.

Berger Chevrolet revived its SS legacy when the fifth-generation Camaro arrived eight years later, this time relying on a supercharged LS3 V-8 bumped up to 550 horsepower. "Our goals for this project were to emphasize the highest possible quality in all areas, choosing proven performance improvements wisely for the best-bang-for-the-buck and using only OEM or better parts throughout," explained dealership president Matthew Berger in November 2009. "The most important aspect of the new Berger SS is that even though the [2010 production car's] performance level has been raised substantially, our version

MODERN MANUFACTURED MUSCLE *(continued)*

maintains the exceptional drivability and reliability of the factory car, but posts even more impressive benchmark numbers." Berger Chevrolet came back with another 550-horsepower Camaro SS, with improved handling, in 2011 before once again returning the reins to the factory guys.

Another familiar name showed up in 2002 after muscle car collector Stefano Bimbi obtained the trademark rights to the Nickey brand and opened Nickey Performance near Chicago in St. Charles, Illinois. The new Nickey works began building "continuation" cars, authentic recreations of those legendary 1960s and 1970s super Chevys, and also offered Stage I, II and III upgrades of the fifth-generation Camaro. It was like living in the past and roaring into the future at the same time.

Perhaps these maximum-muscle machines all came standard with flux capacitors.

The Nickey label, complete with backward "K," was revived in 2002 for a new firm located outside of Chicago, in St. Charles, Illinois. Nickey Performance began offering Stage I, II, and III Camaro upgrades when Chevy's pony car itself was reborn for 2010. *Bob McClurg*

Long known for its Corvette upgrades, Lingenfelter Performance Engineering (based in Brighton, Michigan) didn't overlook Chevrolet's sixth-gen pony car. "The team that developed the 2016 Camaro did a fantastic job," said owner Ken Lingenfelter. "And our engine packages were created to take power, and driving fun, to the next level." Lingenfelter offered two Camaro packages in 2016, one boosting output to 630 horsepower (with a supercharger), the other offering 600 horses via natural aspiration. *courtesy Lingenfelter Performance Engineering*

continued from page 115

Base sticker for a typical V-8 Camaro in 1969 was $2,272. COPO 9560 tacked on another $4,160.15, or just about the amount Gibb figured the entire car would cost. An official price was still in the works when he placed his order; he guessed that $4,900 would be tops. Nearly all the cars had arrived in Illinois before GM sent the bill in March. "I was shocked," Gibb later admitted to *Car Craft's* Dev Annand after setting eyes on the total sticker: $7,269. More than overwhelmed, he somehow managed to send twenty ZL1s back to the Norwood plant, where they were then reluctantly redistributed.

1969 COPO V-8s		
Type	OHV Mk IV L72 big-block[1]	OHV Mk IV ZL1 big-block
Availability	Camaro, Chevelle	Camaro, Corvette
Displacement	427 cid	427 cid
Horsepower	425 at 5,600 rpm	430 at 5,200 rpm
Torque	460 at 4,000 rpm	450 at 4,400 rpm
Compression	11:1	12:1
Fuel delivery	780-cfm Holley 4-barrel	850-cfm Holley double-pumper 4-barrel
Air cleaner	sealed to ZL2 ducted hood	sealed to ZL2 ducted hood
Intake manifold	aluminum, closed-divider	aluminum, open-divider
Ignition	single-point distributor	transistorized
Cooling	heavy-duty Harrison radiator	heavy-duty Harrison radiator
Cylinder block	cast-iron	aluminum w/cast-iron cylinder sleeves
Crankshaft	forged 1053 steel w/4-bolt mains	forged 5410 steel w/4-bolt mains
Connecting rods	forged-steel, ⅜-inch bolts	forged steel, $^{7}/_{16}$-inch bolts
Pistons	aluminum	aluminum
Cam	mechanical	mechanical
Duration	316, intake; 302, exhaust	357, intake; 364, exhaust
Cylinder heads	cast-iron closed-chamber	aluminum open-chamber
Valve sizes	2.19 intake; 1.72 exhaust	2.19 intake; 1.88 exhaust
Valve lift	0.519, intake/exhaust	0.580, intake/exhaust

NOTE: *Bore and stroke for 427-cid V-8 was 4.251 inches x 3.76 inches; valve sizes and valve lift in inches; duration in degrees*

[1]L72 V-8 introduced for Corvette & full-size Chevrolet in1966

Chevy's gnarly M22 Rock Crusher four-speed also was available at extra cost, as were loud, cackling chambered exhausts. Rally wheels, Rally Sport treatment, spoilers, and various other optional upgrades appeared on ZL1 Camaros, too, although such frills were mostly left to the SS/Z28 crowd. Nearly all COPO Camaros sold in 1969 were no-nonsense, easy-to-miss, race-ready machines—Corvette-powered pony cars that GM execs clearly said couldn't be built.

So was Vince Piggins hard of hearing? Not at all; according to Jim Mattison he simply was "damned good at begging for forgiveness instead of asking for permission." Just what you'd expect from someone who never encountered a rule he couldn't break. Or shatter.

COPO COMEBACK

Apparently some things old truly can be new again. Consider the COPO Camaro. In March 2012, Chevrolet announced it would be offering a fifth-generation variation on this theme after unveiling a prototype at Las Vegas's annual Specialty Equipment Market Association show the previous November.

"The COPO Camaro is a proof-of-concept for what a Chevrolet Stock Eliminator entry could look like," explained GM Motorsports chief Jim Campbell in October just prior to the 2011 SEMA event's opening. "And it is a clear indication that Chevrolet intends to homologate Camaro for sportsman drag racing."

Powering that COPO concept was another prototype, a supercharged LSX-based V-8 that displaced 327 cubes. Mentioned as well was an available naturally aspirated 427 derived from Corvette's LS7 small-block, and announced transmission choices originally included a two-speed Powerglide automatic, a three-speed auto, or five-speed manual. Inside was a full roll cage, and out back was a bulletproof solid rear axle in place of the stock Camaro's independent rear suspension. Lightweight racing wheels and radial slicks were present, too, making it more than obvious that Campbell wasn't kidding about this wild horse's intentions.

After more than two thousand rarin'-to-race SEMA attendees expressed their desires to ride this pony hard, one quarter-mile at a time, Chevrolet officials realized they had a no-brainer on their hands—a 2012 COPO simply had to go

2012 COPO Camaro	
Construction	unitized body/frame w/tubular roll cage
Model availability	coupe & convertible
Price	$89,000 (base coupe)
Wheelbase	112.3 inches
Weight	3,175 pounds w/4.0L supercharger V-8, 3,250 pounds w/2.9L supercharger V-8, 3,195 pounds w/naturally aspirated V-8
Front suspension	Strange Engineering adjustable coil-over struts, stock sway bar removed
Rear suspension	solid axle w/4-bar adjustable top links, adjustable Panhard link, double-adjustable Strange Engineering coil-over shocks, and anti-roll bar
Steering	Stock steering gear modified for manual operation (power-assist removed)
Wheels	Bogart Racing lightweight: 15x3.5, front; 15x10, rear
Tires	4.5/28, drag-only, front; 9.0/30 radial slicks, rear
Brakes	Strange Engineering lightweight vented discs w/4-piston lightweight calipers
Engine	5.3L V-8 w/2.9L supercharger, 5.3L V-8 w/4.0L supercharger, or naturally aspirated 7.0L V-8; all recommended to 7,500 rpm maximum
Transmission	ATI Racing Products "Pro Glide" (modified 2-speed Powerglide) w/Hurst Quarterstick shifter featuring built-in neutral safety switch (COPO #68 featured a 5-speed manual, #69 featured a 3-speed automatic)
Rear axle	Strange Engineering 9-inch aluminum center section w/35-spline axles
Production (total)	69 (67 coupes, 2 convertibles)
Production (engine count[1])	43 w/7.0L, 20 w/4.0L supercharged 327; 6 w/2.9L supercharged 327
Production (paint count[2])	32 w/Summit White, 20 w/black, 12 w/ Victory Red, 2 w/Ashen Gray Metallic, 2 w/Silver Ice Metallic, 1 w/Inferno Orange Metallic[3]

[1]Eight customers opted for the "collector's package," which included all three available engines number-matched to each car
[2]Twelve cars were adorned with special COPO graphics
[3]Convertible

"While it was developed strictly for the drag strip, the COPO Camaro is infused with the same performance pedigree that every Camaro shares," said GM Motorsports VP Jim Campbell while introducing this reborn legend in March 2012. "COPO builds off the strengths that have made Camaro the best-selling sports car in America." ***GM***

nto production. "The COPO Camaro is going to shake up drag racing ranks this summer and give Chevy fans a great new reason to cheer on the Bowtie," added Campbell in March 2012 while spreading the good news to drag kings across America. "COPO builds off the strengths that have made Camaro the best-selling sports car in America. And while it was developed strictly for the strip, COPO is infused with the same performance pedigree that every Camaro shares."

But there was a catch. Production was limited to only sixty-nine, the same number of ZL1 Camaros built for 1969. To fairly address the more than three thousand potential buyers who signed up online for the chance to take home a 2012 COPO, Chevrolet turned to a neutral third party to randomly select the lucky sixty-nine, who then were directed to fork over at least $89K for the basic package, which changed slightly compared to the concept. Only one transmission, the race-prepped Powerglide, was offered publically along with three engines: two blown 327s and the non-boosted 427. Also available was a "collector's package" that contained all three engines, each serial-number-match to the car. Five exterior finishes were possible—black, Summit White, Victory Red, Silver Ice Metallic, and Ashen Gray Metallic—plus an optional graphics treatment similar to that seen in Vegas on the concept vehicle.

The first sixty-seven 2012 COPOs were coupes. Number sixty-eight was a convertible not meant for public sale, while the last of the run was an Inferno Orange ragtop sent to auction to benefit the American Heart Association. It

COPO COMEBACK *(continued)*

The last two of sixty-nine COPO Camaros built in 2012 were convertibles, with the final one (shown here) auctioned off for charity in January 2013. *GM*

sold for $400,000 in January 2013. Number sixty-eight featured a five-speed manual transmission, while number sixty-nine was fitted with a three-speed automatic.

Another sixty-nine COPO Camaros were released in 2013, and again in 2014 and 2015. Changes for 2013 included an available G-Force G101A four-speed manual transmission (along with the rollover Powerglide) and three new naturally aspirated engines, a 350-horsepower 350, 500-horsepower 396 and 435-horsepower 427, all fitted with Holley HP electronic fuel injection. Revised front springs and a radiator-integrated trans cooler also were included this year. The naturally aspirated 396 and 427 V-8s were revised slightly for 2014 and were joined by a supercharged 350, rated at 530 horsepower. Updated graphics appeared in 2015 as the fifth-gen Camaro rolled into the sunset.

Jim Campbell announced Chevy's next-generation strip-ready pony in November 2015. "With the new Gen Six Camaro as its foundation, the 2016 COPO begins the next chapter of an historic racing legacy." Indeed, benefitting from the lighter, stronger sixth-gen platform, the latest COP Camaro promised even more success at the drags. A roll cage again was included inside, as was a solid axle out back. Three engines were offered: a supercharged 350, unblown 427, and a new LT-based 376. The only transmission choice this time was a race-prepped three-speed Turbo 400 automatic.

Eager buyers again had to register online for the chance to score one of 2016's sixty-nine COPOs. How'd that third-party selection process work for you?

2012 COPO Camaro Engines

	5.3L	5.3L	7.0L
Type	OHV V-8 w/2.9L supercharger	OHV V-8 w/4.0L supercharger	naturally aspirated OHV V-8
CID	327	327	427
Bore x stroke	4.065 x 3.150	4.065 x 3.150	4.125 x 4.00
Horsepower[1]	500	550	425
Compression[2]	10.0:1	10.0:1	13.0:1
Fuel delivery	2.9L twin-screw Whipple Industries supercharger & Whipple billet aluminum throttle body w/cable-actuation	4.9L twin-screw Whipple Industries supercharger & Whipple billet aluminum throttle body w/cable-actuation	cable-actuated Whipple Industries billet aluminum throttle body on Chevrolet Performance/Holley "Hi Ram" intake manifold
Ignition	Delco MEFI 5A processor w/LS7 coils	Delco MEFI 5A processor w/LS7 coils	Delco MEFI 5A processor w/LS7 coils
Cooling	Chevrolet Performance LS3 water pump	Chevrolet Performance LS3 water pump	Meziere Enterprises billet electric water pump
Oiling	internal pump w/cast-aluminum deep-sump pan	internal pump w/cast-aluminum deep-sump pan	internal pump w/cast-aluminum deep-sump pan
Cylinder block	cast-iron Chevrolet Performance LSX w/steel main bearing caps	cast-iron Chevrolet Performance LSX w/steel main bearing caps	cast-iron Chevrolet Performance LSX w/steel main bearing caps
Crankshaft	Callies 4340 "Dragonslayer"	Callies 4340 "Dragonslayer"	Callies 5140 "Compstar"
Connecting rods	H-beam Callies 4340 "Ultra Rods"	H-beam Callies 4340 "Ultra Rods"	H-beam Callies 4340 "Compstar"
Pistons	Mahle forged 2618 alloy	Mahle forged 2618 alloy	Mahle forged 2618 alloy
Cam	Comp Cams steel billet; 246/254 degrees duration, intake/exhaust	Comp Cams steel billet; 246/254 degrees duration, intake/exhaust	Chevrolet Performance steel billet; 233/276 degrees duration, intake/exhaust
Lifters	Chevrolet Performance "Ceramic Ball" hydraulic rollers	Chevrolet Performance "Ceramic Ball" hydraulic rollers	Chevrolet Performance "Ceramic Ball" hydraulic rollers
Cylinder heads	aluminum LSX-based LS7	aluminum LSX-based LS7	aluminum LS7-based
Rocker arms	LS7 rollers, 1.8:1 ratio	LS7 rollers, 1.8:1 ratio	LS7 rollers, 1.8:1 ratio
Valves	Del West titanium intakes, lightweight sodium-filled exhausts	Del West titanium intakes, lightweight sodium-filled exhausts	Del West titanium intakes, lightweight sodium-filled exhausts
Valve sizes	2.205/1.615, intake/exhaust	2.205/1.615, intake/exhaust	2.205/1.615, intake/exhaust
Valve lift	0.630 intake/exhaust	0.630 intake/exhaust	0.630 intake/exhaust
Exhaust	American Racing stainless-steel tubular headers	American Racing stainless-steel tubular headers	American Racing stainless-steel tubular headers

CID is cubic-inch displacement; bore & stroke, valve sizes & valve lift in inches

[1]NHRA rating

[2]Nominal

1970
CAMARO Z28

6

A MORE SOPHISTICATED CAMARO

Second generation: 1970–1981

Resting on laurels has never been a good idea in Detroit. Hence, Henry Haga's team was typically busy sculpting a new and improved F-car even as Chevrolet's original Camaros were rolling down the Van Nuys and Norwood lines in 1966. And, in equally familiar fashion, the supposedly secret word about this future product just couldn't help but escape into the wild well ahead of time. By 1969, pony lovers from California to Ohio (as well as all points east) were eagerly awaiting an expected rebirth, surely due that fall.

Then a funny thing happened on the way to the delivery room. When Chevrolet's spanking-new models met the world in September, a fresh-faced 1970 Camaro failed to appear. The wildly popular 1969 version remained on sale while Chevrolet readied a dramatically redesigned Camaro to release later in the model year.

GM execs, meanwhile, continued teasing eager buyers. "We will give an entirely new direction to this market," claimed Chevy general manager John DeLorean in a January 1970 *Motor Trend* interview. The upcoming next-gen Camaro, in his words, would be "so sensational that I think we will more than make up for lost ground."

(opposite)
Chevrolet's second-generation Camaro was a late entry for the 1970 model year. The breed's performance flagship, Z28, rolled on, but only until 1974. *GM*

Indeed, almost all was forgiven when a born-again Camaro finally did show in February 1970. "Worth the wait" was *Car Life's* description for a thoroughly modernized machine that stood "a whole light year away from last year's." Low and wide, this sleek sport coupe was commonly praised for the way its fancy facade and upgraded feel masked its affordable, comparatively compact nature. Engineered updates included standard front disc brakes and safety-conscious side impact beams welded into the doors.

"Innovations in design and engineering make the new Camaro different from any car now offered," announced DeLorean during a press conference held on February 13—a Friday, no less. "Its expensive-looking body and highly improved handling, ride and sound insulation create a completely new car." According to *Detroit News* writer Bob Irvin, the so-called 1970½ Camaro was "a beautiful looking automobile, one that makes Mustang and Barracuda seem like last year's models." In the words of GM styling boss Bill Mitchell, the second-generation was "a designer's design," while the first was "designed by committee."

FROM THE HORSE'S MOUTH VI

"I've always loved the 1968 Camaro, but for some reason, it was the second-gen car I'd find myself sketching during high school math. It was a radical departure from the first-gen. For the first time, it was built on its own dedicated architecture, which gave the design team the freedom to create a pure expression. What that team created was a powerful expression of American muscle, influenced by a European grand-touring aesthetic. There was simply nothing else like it. The second-gen car is pure Camaro, with a dramatic proportion and lean, muscular form. You won't confuse it with the first generation, but it is unmistakably a Camaro from every angle."

GM Design vice president, China
Ken Parkinson, February 2015

While he prefers the second-generation models aesthetically, Ken Parkinson owns a 1968 SS/RS, a small-block model with a four-speed stick. He is now vice president of GM Design, China. *GM*

Roots of that design ran back to June 1966, when second-gen sketches (albeit far-fetched ones) first started showing up in Haga's office. An initial clay mockup appeared in August, as did (on paper, that is) a two-seat convertible not long afterward. Meanwhile, in Bill Porter's Pontiac studio, a roofline was taking shape that would carry over almost unchanged (save for a proposed wraparound rear window) into production for both Firebird and Camaro. In the end, GM's F-car duo shared identical upper structures mated to division-specific lower sheet metal. Thus, rear-quarter glass was deleted in both cases, saving about $18 per car.

High costs, versus minimal returns, also torpedoed all topless considerations. Only one body, that swoopy sport coupe (referred to as a "fastback" by some), was offered during the second-generation run. According to Haga, inspiration for the 1970 Camaro's front end was supplied by Jaguar, while other influences also came from across the Atlantic as well.

1970 Camaro	
Construction	unitized body w/front subframe
Model availability	sport coupe
Price	$2,749, 6-cyl.; $2,839, V-8
Wheelbase	108 inches
Curb weight (pounds)	3,166, 6-cyl.; 3,278, V-8
Suspension	independent short/long-arm (SLA) w/coil springs & 0.6875-inch stabilizer bar, front; solid-axle w/multi-leaf springs, rear
Steering	24:1 manual recirculating-ball, std; variable-rate power assist, optional
Wheels	14x6 stamped-steel
Tires	E78-14
Brakes	11-inch discs, front; 9.5-inch drums, rear; power-assist, optional
Base engines	155-hp Turbo Thrift 230 inline-six; 200-hp Turbo Fire 307 V-8
Transmission	3-speed manual, std; Powerglide automatic, optional (w/6-cyl. & 307 V-8), 4-speed manual, optional (w/350 V-8s), TH 350 automatic, optional (w/Turbo Fire V-8s); TH 400 automatic, optional (w/Turbo Jet V-8s)
Axle ratio	3.08:1, std w/3-speed manual; 2.73:1 included w/Powerglide; 3.36:1 included w/L65 V-8 & 4-speed; 3.31:1 included with L48 V-8 & 4-speed

As *Car and Driver* explained it, "there is good reason to believe that Camaro and Firebird are the leading edge of a new trend in Detroit styling. Europeans

Various factors delayed the new-for-1970 Camaro's debut, leaving Chevrolet officials little choice but to offer rebadged 1969 leftovers. Notice the front license plate. *GM*

Early proposals for the second-generation Camaro were wild, to say the least, but the design quickly settled down and was taking familiar shape by August 1968. *GM*

have popularized the concept of a strong, simple shape with extreme tumblehome and tuckunder that doesn't need stick-on ornamentation to make it work. Camaro is certainly of this school."

Beauty beneath that strong shape more or less carried over from 1969. Save for the latest Z28's Corvette-sourced 350 V-8, powertrain choices rolled over unchanged (though with lower compression and less power), as did the basic unitized body-frame structure. Engineers even considered keeping those Mono Plates in back before conceding that multi-leaf springs (again with staggered shocks) were a must for 1970's extra weight. They also mildly revised the foundation (both floorpan and subframe) to increase rigidity and reworked the front suspension to enhance handling and ride. Most notably, steering linkage was relocated from behind the wheel axis to in front, where it contributed to understeer conditions, something casual drivers can deal with far easier than the oversteer created by 1969's "rear-steer" design.

According to *Hot Rod's* Steve Kelly, the sum of these parts equaled a vehicle that "extends the original concept closer towards the sports side of 'sports/compact.'" Raising the praise bar even higher, *Road & Track* claimed "the new Camaro represents what we think is the first serious effort since the 1963

While Camaro has never offered three body style choices like Ford's Mustang did from 1964 to 1973, consideration was given to various proposals, included a two-seater and a station wagon of sorts called a "Kammback." Kammback (or "Nomad") concepts were modeled during both first- and second-generation development, with the latter pictured here. *GM*

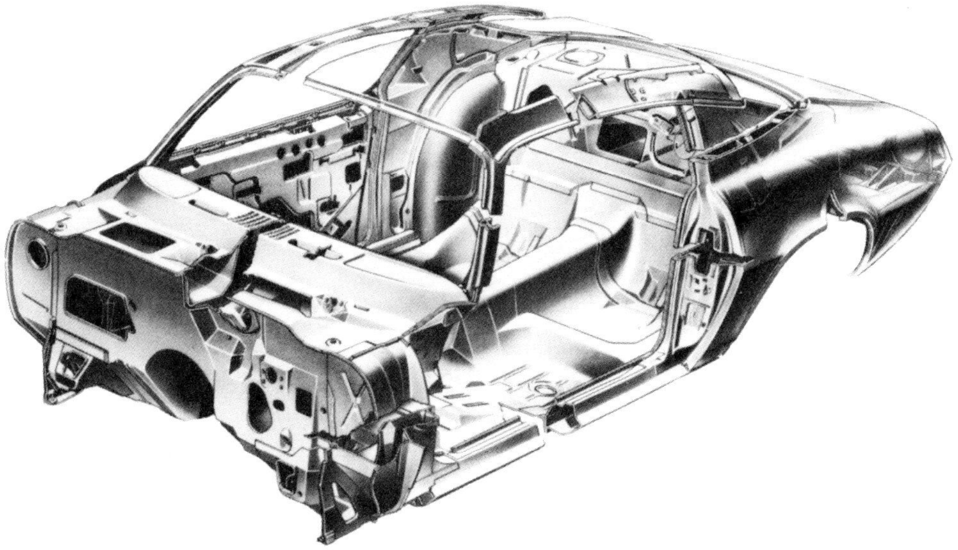

Semi-unitized construction (incorporating a front subframe) remained the norm for the second-gen Camaro. Sill moldings were widened and door openings were enlarged to improve rear seat entrance/exit. *GM*

Corvette to create a real American GT. A tremendous improvement, [it] puts [Chevrolet's] Pony Car in a new class."

The rendition *Road & Track* referred to was a small-block Super Sport, which rolled over into 1970 along with the aforementioned Z28 and ever-popular Z22 Rally Sport. Still available in Z28 and Z27 SS packages, 1970's RS deal was less prominent now that hideaway headlights were no longer possible. But it still stood out thanks to two small bumpers found up front in place of the standard full-width unit. Various bright accents and black-out treatments, along with specific identification, completed the second-gen Z22 option.

With hideaway headlights no longer possible, 1970's Rally Sport had to rely on a pair of small front bumpers to set itself apart. Unfortunately, this minimally defensive design would help spell an end for the original RS image three years later after federal safety standards began mandating that bumpers actually serve a purpose. *GM*

Chevy's legendary Z-car managed to survive until 1975 before fading away along with the great American muscle car genre. Its return in 1977 represented a minor miracle. Not so lucky was Camaro SS, which retired in 1972, along with the breed's last big-block V-8. An F-body Super Sport revival wouldn't arrive until twenty-four years later.

In the Super Sport's place was 1973's Type LT, a classy V-8 model that came standard with extra body insulation, variable-rate power steering, hidden windshield wipers, sport mirrors, full instrumentation, and 14x7 Rally wheels. First offered in 1970, hidden wipers (RPO C24) were included on all models beginning in 1977, the same year Chevy's straight-six became the Type LT's base engine.

Available in combination with Z28 and/or RS gear, the Type LT carried on as an individual model up through 1978 before it was also replaced, in this case by the plush Berlinetta. Various bright trim bits were standard Berlinetta fare, as

Chevrolet's Sport Cabriolet (upper left), introduced in 1927 was a closed car that looked like a convertible thanks to its canvas-covered top. No topless Camaros were offered during the second-gen run, leaving models like the relatively affordable, yet decently flashy Rally Sport coupe (pictured here) to turn customers' heads in 1972. *GM*

1970–1981 Camaro Engines					
Year	**RPO**	**Displacement (CID)**	**Horsepower**	**Torque**	**CR**
1970	std 6-cyl.	250	155 at 4,200 rpm	235 at 1,600 rpm	8.5:1
	std V-8	307	200 at 4,600 rpm	300 at 2,400 rpm	9.0:1
	L65 V-8	350	250 at 4,800 rpm	345 at 2,800 rpm	9.0:1
	L48 V-8	350	300 at 4,800 rpm	380 at 2,200 rpm	10.25:1
1971	std 6-cyl.	250	145 at 4,200 rpm	230 at 1,600 rpm	8.5:1
	std V-8	307	200 at 4,600 rpm	300 at 2,400 rpm	8.5:1
	L65 V-8	350	245 at 4,800 rpm	350 at 2,800 rpm	8.5:1
	L48 V-8	350	270 at 4,800 rpm	360 at 3,200 rpm	8.5:1
1972	std 6-cyl.	250	110 at 3,600 rpm	185 at 1,600 rpm	8.5:1
	std V-8	307	130 at 4,000 rpm	230 at 2,400 rpm	8.5:1
	L65 V-8	350	165 at 4,000 rpm	280 at 2,400 rpm	8.5:1
	L48 V-8	350	200 at 4,400 rpm	300 at 2,800 rpm	8.5:1
1973	std 6-cyl.	250	100 at 3,600 rpm	175 at 1,600 rpm	8.5:1
	std V-8	307	115 at 3,600 rpm	205 at 2,000 rpm	8.5:1
	L65 V-8	350	145 at 4,000 rpm	255 at 2,400 rpm	8.5:1
	L48 V-8	350	175 at 4,000 rpm	260 at 2,800 rpm	8.5:1
1974	std 6-cyl.	250	110 at 3,600 rpm	175 at 1,600 rpm	8.5:1
	L65 V-8[1]	350	145 at 3,800 rpm	250 at 3,200 rpm	8.5:1
	LM1 V-8	350	160 at 3,800 rpm	250 at 2,400 rpm	8.5:1
	L48 V-8[2]	350	185 at 4,000 rpm	270 at 2,600 rpm	8.5:1
	[1]standard V-8 in all states except California *[2]available all states*				
1975	std 6-cyl.	250	105 at 3,800 rpm	185 at 1,200 rpm	8.25:1
	L65 V-8[3]	350	145 at 3,800 rpm	250 at 2,800 rpm	8.5:1
	LM1 V-8[4]	350	155 at 3,800 rpm	250 at 2,400 rpm	8.5:1
	[3]standard V-8 in all states except California *[4]available in all states*				
1976	std 6-cyl.	250	105 at 3,800 rpm	185 at 1,200 rpm	8.25:1
	LG3 V-8	305	140 at 3,800 rpm	245 at 2,000 rpm	8.5:1
	LM1 V-8	350	165 at 3,800 rpm	260 at 2,400 rpm	8.5:1
1977	L22 6-cyl.[5]	250	110 at 3,800 rpm	195 at 1,600 rpm	8.3:1
	L22 6-cyl.[6]	250	90 at 3,600 rpm	180 at 1,600 rpm	8.3:1
	LG3 V-8	305	145 at 3,800 rpm	245 at 2,400 rpm	8.5:1
	LG3 V-8[6]	305	135 at 3,800 rpm	240 at 2,000 rpm	8.5:1
	LM1 V-8	350	170 at 3,800 rpm	270 at 2,400 rpm	8.5:1
	LM1 V-8[6]	350	160 at 3,800 rpm	260 at 2,400 rpm	8.5:1
	LM1 V-8[7]	350	185 at 4,000 rpm	280 at 2,400 rpm	8.5:1
	[5]base engine except in California [6]available in California *[7]Z28 V-8 (rated at 175 horsepower in California)*				
1978	L22 6-cyl.[8]	250	110 at 3,800 rpm	190 at 1,600 rpm	8.1:1
	LG3 V-8[9,10]	305	145 at 3,800 rpm	245 at 2,400 rpm	8.4:1
	LM1 V-8[9,11]	350	170 at 3,800 rpm	270 at 2,400 rpm	8.2:1
	LM1 V-8[9,12]	350	185 at 4,000 rpm	280 at 2,400 rpm	8.2:1
	[8]base engine, all states (rated at 90 horsepower in California & high altitudes [9]optional, all states *[10]rated at 135 horsepower in California & high altitudes [11]rated at 160 horsepower in California & high altitudes* *[12]Z28 V-8 (rated at 175 horsepower in California & high altitudes)*				

1970–1981 Camaro Engines *(continued)*					
Year	**RPO**	**Displacement (CID)**	**Horsepower**	**Torque**	**CR**
1979	L22 6-cyl.[13]	250 cid	115 at 3,800 rpm	200 at 1,600rpm	8.0:1[14]
	LG3 V-8[15, 16]	305 cid	130 at 3,200 rpm	245 at 2,000 rpm	8.4:1
	LM1 V-8[15, 17]	350 cid	170 at 3,800 rpm	270 at 2,400 rpm	8.2:1
	LM1 V-8[15, 18]	350 cid	175 at 4,000 rpm	270 at 2,400 rpm	8.2:1
	[13]base engine, all states (rated at 90 horsepower in California) [14]8.2:1 in California [15]optional, all states [16]rated at 125 horsepower in California [17]rated at165 horsepower in California & high altitudes [18]Z28 V-8 (rated at 170 horsepower in California & high altitudes)				
1980	LC3 V-6[19]	229 cid	115 at 4,000 rpm	175 at 2,000 rpm	8.6:1
	LD5 V-6[20]	231 cid	110 at 3,800 rpm	190 at 1,600 rpm	8.0:1
	L39 V-8[21]	267 cid	120 at 3,600 rpm	215 at 2,000 rpm	8.3:1
	LG4 V-8[22]	305 cid	155 at 4,000 rpm	240 at 1,600 rpm	8.6:1
	LG4 V-8[23]	305 cid	165 at 4,000 rpm	280 at 2,400 rpm	8.6:1
	LM1 V-8[24]	350 cid	190 at 4,200 rpm	280 at 2,400 rpm	8.2:1
	[19]base engine, all states except California [20]California only, produced by Buick [21]optional all states, except California [22]available all states (California torque rating: 230 at 2,400 rpm); n/a w/Z28 [23]Z28 V-8 in California [24]optional all states except California; Z28 V-8 in all states except California				
1981	LC3 V-6[25]	229 cid	110 at 4,200 rpm	170 at 2,000 rpm	8.6:1
	LD5 V-6[26]	231 cid	110 at 3,800 rpm	190 at 1,600 rpm	8.0:1
	L39 V-8[27]	267 cid	115 at 4,000 rpm	200 at 2,400 rpm	8.3:1
	LG4 V-8[28]	305 cid	150 at 3,800 rpm	240 at 2,400 pm	8.6:1
	LG4 V-8[29]	305 cid	165 at 4,000 rpm	245 at 2,400 rpm	8.6:1
	LM1 V-8[30]	350 cid	175 at 4,000 rpm	275 at 2,400 rpm	8.2:1
	[25]base engine, all states except California [26]California only, produced by Buick [27]optional all states, except California [28]optional all states; n/a w/Z28 [29]Z28 V-8 w/manual transmission in all states [30]Z28 V-8 w/automatic transmission in all state				

NOTE 1: *CR is compression ratio; see Ghapter 3 for 1970-72 Super Sports V-8s; see Chapter 4 for 1970-74 Z28 V-8s*
NOTE 2: *1970 & 1971 outputs figures are gross numbers; 1972-77 are net*

(opposite)
Those aforementioned federal safety standards helped inspire a notable facelift for 1974's Camaro. A larger fuel tank also was new this year, as were optional radial tires. *GM*

were color-keyed wheels and sport mirrors, pinstriping, and an argent rear panel appliqué. Like the last LT, the first Belinetta also featured six cylinders in base form. It remained in the Camaro fold up through 1986.

An F-car staple since the get-go, Camaro's original Rally Sport package was dropped after 1973 but then reappeared midyear in 1975, now coded Z85. Most notable among the Z85 features was upper-body black-out paint, which could be applied to both base coupes and Type LT models. "Must witness to appreciate" also applied in each case.

"Whoever started this paint-the-hood-black craze must have seen *Twelve O'Clock High* too many times," reported Stephan Wilkinson in the August 1975 issue of *Car and Driver*. "But the hype apparently works. Though it hardly serves the kind of anti-dazzle function it [did] on aircraft, the paint gets more than its share of glances on the highway and attracts gas-station attendants the way a Carrera's foot-high gold lettering never will."

Stronger side impact beams and optional power windows were new for 1973, as was the Type LT, with its upscale interior. An individual model, not an options group, the Type LT could be adorned further with Z28 and/or RS packages. ***Jim Schild***

1976 LG3 V-8	
Type	OHV Gen-I small-block
Displacement	305 cubic inches (5.0 liters)
Bore	3.736 inches
Stroke	3.48 inches
Horsepower	140 at 3,400 rpm
Torque	245 at 2,000 rpm
Compression	8.5:1
Fuel delivery	Rochester 2-barrel carburetor
Air cleaner	15.48-inch single-snorkel
Ignition	HEI distributor
Cooling	18-inch 4-blade fan
Cylinder block	cast-iron w/5 main bearings
Crankshaft	cast-nodular iron w/2.4493-inch main journals & 2.0988-inch rod journals; 2-bolt main bearing caps
Connecting rods	forged-steel
Pistons	cast-aluminum
Cam	hydraulic w/272 degrees duration on intake, 288 degrees on exhaust
Rocker ratio	1.50:1
Cylinder heads	cast-iron w/ball-stud rockers & 5.13-cubic-inch combustion-chamber volume
Valve sizes (inches)	1.72, intake; 1.50, exhaust
Valve lift (inches)	0.3727, intake; 0.410, exhaust
Exhaust system	cast-iron rear-dump manifolds w/2-inch pipes, single catalytic converter, single muffler

In 1978, Camaro's Rally Sport treatment was transformed from an options group to full model status, with both a base sport coupe RS and Type LT RS offered. As in 1977, the Rally Sport paint/striping exterior could not be combined with 1978's Z28, and the same exclusion applied to 1979's new Berlinetta. A stand-alone Camaro RS appeared once more in 1980 before rolling off into the archives.

As for other standard/optional feature timelines, 1972 was the last year for an available Powerglide automatic, 1980 the last for a big-boy spare tire, and 1981 the last for a three-speed stick. Originally available in 1968 and 1969, the toy-like N65 space-saver spare reappeared on 1973's RPO list then became a standard Camaro feature, along with power brakes, for all 1981 models. Power-assisted stoppers had been included with V-8 installations since 1976, as had power steering dating back to 1974. A new option in 1969, variable-ratio power steering joined the standard equipment list for all models in 1975. Additional optional debuts of note included power windows in 1973, radial tires in 1974, power door locks in 1975, and T-tops in 1978.

In Z28 ranks, aluminum wheels were new in 1978, as was a front air dam and functional fender

A wider, wraparound rear window appeared in 1975 in response to complaints of poor rearward vision. But making even bigger news was the reborn Rally Sport, which featured a trendy black-out paint treatment for the hood, roof, and upper doors. *GM*

Available as an option (RPO C24) since 1970, hidden windshield wipers became standard on all models in 1977. An LT coupe (priced at $4,478 with the base six-cylinder) appears here. *GM*

louvers the next year. Also fully functional, a rear-facing hood scoop joined Z28's standard equipment list in 1980, followed by a computer-controlled torque converter for 1981's auto-trans models.

In underhood affairs, 1973 was the last year for the 307-cid base V-8 and 1974 was the first for the returning LM1 350 option, net-rated at 160 horsepower. A 145-horsepower 350 served as Camaro's standard V-8 in 1975 then was supplanted by a 140-horsepower 305-cid small-block in 1976, the last year for the six-cylinder/V-8 model line segregation. The yeoman 250-cid six was standard for 1977's basic coupe and Type LT, while the reincarnated Z28 was treated to an exclusive 350, rated at 185 horsepower. Optional from 1977 through 1981 for non-Z models were the LM1 350 and 305 small-blocks, joined in 1980 by an available 267-cid V-8, RPO L39.

New for all Camaro engines in 1975 was High Energy Ignition, an electronic Delco-Remy distributor first seen beneath Z28 hoods the previous year. HEI did away with maintenance-intensive breaker points (and the required resistor circuit) in favor of a worry-free control module that also increased voltage. Leaner fuel mixtures were allowed, too, as HEI use meant spark plug gaps were widened.

Camaro chief engineer Tom Zimmer hovers proudly over the various hot parts that made Chevrolet's revived Z28 a midyear sensation in 1977. *GM*

1980 L39 V-8	
Type	OHV Gen-I small-block
Displacement	267 cubic inches (4.4 liters)
Bore	3.50 inches
Stroke	3.48 inches
Horsepower	120 at 3,600 rpm
Torque	215 at 2,00 rpm
Compression	8.3:1
Fuel delivery	Rochester 2-barrel carburetor
Air cleaner	12-inch single-snorkel
Ignition	HEI distributor
Cooling	18-inch 4-blade fan (5-blade w/3.08:1 axle)
Cylinder block	cast-iron w/5 main bearings
Crankshaft	cast-nodular-iron w/2-bolt main bearing caps
Connecting rods	forged-steel
Pistons	cast-aluminum
Cam	hydraulic w/272 degrees duration on intake, 288 degrees on exhaust
Rocker ratio	1.50:1
Cylinder heads	cast-iron w/bull-stud rockers & 4.58-cubic-inch combustion-chamber volumes
Valve sizes (inches)	1.72, intake; 1.38, exhaust
Valve lift (inches)	0.357, intake; 0.390, exhaust
Exhaust system	cast-iron manifolds w/2-inch pipes, single catalytic converter, single muffler

More modernization came in 1980 when Camaro's standard inline-six was replaced by a lighter V-6, a 229-cid engine rated at 115 horsepower. Sourced from Buick, a second V-6 displacing 231 cubes was installed in California-delivered cars. Both were advertised at 3.8 liters, and the West Cost six was covered by printed disclaimers explaining that "Chevrolets are equipped with GM-built engines produced by various divisions." Called in some quarters a "three-quarter small-block," that Bowtie-built V-6 was the forerunner to the division's popular Vortec truck engines.

In more general news, a major facelift appeared for 1974 as part of Chevrolet's compliance with federal safety standards that required bumpers to actually be crash proof (at slow speeds), as opposed to simply looking the part. Extra reinforcing appeared initially in 1973 to meet these requirements, and these modifications even worked to Washington's satisfaction for the Rally Sport's two front "bumperettes." But this band-aid approach didn't stand up long, explaining why the RS didn't return for 1974. New that year were cushioned aluminum bumpers at both ends, a fiberglass front fascia, and

Another facelift for the second-gen Camaro appeared in 1978 and was followed in 1979 by the new Berlinetta (shown here), which superseded the Type LT. Like 1978's LT, the upscale Berlinetta featured six-cylinder power in base form. *GM*

Big news for 1980's base sport coupe, Rally Sport (shown here) and Berlinetta included a modern V-6 in place of the archaic inline-six installed from 1967 to 1979. *GM*

revised taillights. Also included in this functional makeover were seven extra inches of length and more than two hundred additional pounds.

While most designs failed miserably (aesthetically speaking) during the "big bumper" era's early years, the Camaro crew's work arguably demonstrated that they apparently knew something their rivals didn't. According to *Small Cars'* Mike Uno, Chevy's freshened F-car was "the most tastefully styled of the '74s."

A functional hood scoop was new for 1980's Z28, powered by a 190-horsepower 350 V-8 in all states except California. Standard there was a 305 small-block rated at 165 horsepower. *GM*

1980 LC3 V-6	
Type	OHV V-6
Displacement	229 cubic inches (3.8 liters)
Bore	3.736 inches
Stroke	3.48 inches
Horsepower	115 at 4,000 rpm
Torque	175 at 2,000 rpm
Compression	8.6:1
Fuel delivery	Rochester 2-barrel carburetor
Air cleaner	12-inch single-snorkel
Ignition	HEI distributor
Cooling	18-inch 4-blade fan
Cylinder block	cast-iron w/4 main bearings
Crankshaft	cast-nodular-iron w/2.4486-inch main bearing journals; 2-bolt main bearing caps
Connecting rods	forged-steel
Pistons	cast-aluminum
Cam	hydraulic w/300 degrees duration on intake, 310 degrees on exhaust
Rocker ratio	1.50:1
Cylinder heads	cast-iron w/5.013-cubic-inch combustion-chamber volumes
Valve sizes (inches)	1.84, intake; 1.50, exhaust
Valve lift (inches)	0.373, intake; 0.410, exhaust
Exhaust system	cast-iron manifolds w/2-inch pipes, single catalytic converter, single muffler

OPENING THE GATE

Pete Estes's September 1966 prediction that Camaro would someday outsell Mustang finally came true in 1977. But which arrived first?—the former's rise or latter's fall? Lee Iacocca never liked the way his baby ballooned after 1966, and he wasn't alone by the time an even larger pony appeared for 1971 to thoroughly disgust many loyal customers. As early as November 1969, he started suggesting bringing Mustang back in line with his original ideal, a plea endorsed by the success of Dearborn's new compact, Maverick, in 1970. Italy's famed design studio, Ghia, then was responsible for creating the first "Mustang II" prototype. Based on Pinto's platform, Iacocca's "little jewel" was officially introduced on August 28, 1973. Wheelbase was a scant 96.2 inches, down 13 from its 1973 predecessor. Mustang II also was 4 inches skinnier, 14 inches shorter, and some 300 pounds lighter. A 2.3 liter four-cylinder was standard, a 2.8 liter V-6 optional. Proving that less sometimes can be more, the truly small Mustang II set a sales pace nearly as quick as its ancestor had ten years before. More than 385,000 buyers took home Mustang IIs in 1974, a good number of those perhaps jumping on the bandwagon after *Motor Trend* named Ford's toy pony its "Car of the Year" in January 1974. In *Motor Trend*'s words, this new breed represented "a total departure from the fat old horse of the recent past." Others were not so quick to praise. *Car Craft* called the Mach 1 rendition "regrettably underpowered" and predicted that Ford surely would offer a V-8 version, and soon. But while an optional V-8 did arrive in 1975, sales still fell by more than 50 percent. Many buyers simply didn't like those cramped quarters, and it seemed soaring gas prices could only compromise a Mustanger's standards so far. Sagging demand then signaled the need for yet another redesign, which debuted for 1979—and put Ford back on top in Detroit's horse race.

Lee Iacocca's "Little Jewel" was a big hit in 1974, scoring *Motor Trend's* Car of the Year trophy and bringing buyers galloping into Ford showrooms much like they'd done in 1964. But the gleam quickly wore off, allowing Camaro to take the pony car sales lead for the first time in 1977. *GM*

Power brakes and a space-saver spare became standard equipment in 1981, which also was the last year for three-speed manual transmissions. A computer-controlled torque converter was new for this year's automatic-transmission Z28.

"Camaro, along with the very similarly styled Firebird, is about the only automobile that displays an imaginative approach to blending in the federally mandated impact-absorbing front and rear bumpers," he continued. "Rather than have the bumpers appear as some sort of 'ramming prows,' that have been merely tacked onto the car, Chevrolet designers have managed to extend the already clean lines of the Camaro into a very subtle and attractive nose. At the rear of the car, the bumper has been affixed so as to appear a logical extension for the Camaro's gently sloping rearward lines."

An enlarged wraparound rear window (first seen, as mentioned, in Pontiac's design studio late in 1966) was added in 1975 in response to complaints about poor visibility, and another restyle followed three years later to take the crash-resistant bumper theme to even greater aesthetic heights. Body-colored urethane front and rear fascias disguised the impact-absorbing equipment and were joined by a revised grille and another set of updated taillights.

1970–1981 Production Figures

	Body Style	6-cyl.	V-8	Type LT	Z28	Total
1970	sport coupe	12,578	112,323	n/a	NOTE 1	**124,901**[1]
	[1]included 27,136 w/Z22 Rally Sport Package					
1971	sport coupe	11,178	103,452	n/a	NOTE 1	**114,630**[2]
	[2]included 18,404 w/Z22 Rally Sport Package					
1972	sport coupe	4,821	63,830	n/a	NOTE 1	**68,651**[3]
	[3]included 11,364 w/Z22 Rally Sport Package					
1973	sport coupe	3,614	60,810	32,327	NOTE 1	**96,751**[4]
	[4]included 16,133 w/Z22 Rally Sport Package					
1974	sport coupe	22,210	79,835	48,963	NOTE 1	**151,008**
1975	sport coupe	29,749	76,178	39,843	NOTE 1	**145,770**[5]
	[5]included 7,000 w/Z85 Rally Sport Package					
1976	sport coupe	38,047	92,491	52,421	n/a	**182,959**[6]
	[6]included 15,855 w/Z85 Rally Sport Package					
1977	sport coupe	131,717		72,787	14,349	**218,853**[7]
	[7]included 17,026 with Z85 Rally Sport Package					

		Base	RS	Type LT	LT RS	Z28	Total
1978	sport coupe	134,491	11,902	65,635	5,696	54,907	**272,631**
				Berlinetta			
1979	sport coupe	111,357	19,101	67,236		84,877	**282,571**
1980	sport coupe	68,174	12,015	26,679		45,137	**152,005**
1981	sport coupe	62,614	n/a	20,253		43,272	**126,139**

NOTE 1: *For 1970–1975 Z28 numbers, see Chapter 4* ***NOTE 2:*** *For 1970–1972 Super Sport numbers, see Chapter 3*
NOTE 3: *Type LT was V-8 only for 1973–1976* ***NOTE 4:*** *A 6-cylinder was standard for 1977's sport coupe & Type LT; a V-8 was optional for both—production breakdown by engine was 31,389 sixes, 187,464 V-8s*
NOTE 5: *A 6-cylinder was standard for 1978's sport coupe, Rally Sport, Type LT, & Type LT Rally Sport; a V-8, optional—production breakdown by engine was 36,982 sixes, 235,649 V-8s* ***NOTE 6:*** *A 6-cylinder was standard for 1979's sport coupe, Rally Sport, &Berlinetta; a V-8, optional—production breakdown by engine was 21,913 sixes, 260,658 V-8s*
NOTE 7: *A V-6 was standard for 1980's sport coupe, Rally Sport &Berlinetta; a V-8, optional—production breakdown by engine was 51,104 sixes, 100,901 V-8s* ***NOTE 8:*** *A V-6 was standard for 1981's sport coupe &Berlinetta, a V-8 optional—production breakdown by engine was 52,004 sixes, 74,135 V-8s*

"Camaro bows into 1978 with a new costume and full-fledged star status in the Chevrolet cast," began *Car and Driver*'s Don Sherman in a salute to this year's Z. "While the 1977 Z28 look was little more than paint and tape make-up over a base Camaro, new tooling money had paid for the most thorough facelift of this old tragedian since the great redesign of 1970. Aluminum bumpers have given way to the soft look. Underneath a smooth mask of flexible urethane, this new Camaro packs front and rear aluminum impact rams padded with molded plastic cushions. Weight savings [amount to] zero in front but almost 50 pounds in back. The execution is identical to Pontiac's Firebird and very similar to Porsche's 928, each living proof that bumper cars can be beautiful."

But transforming that beauty back into a beast would have to wait for the next generation, delivered in 1982.

1982 CAMARO Z28

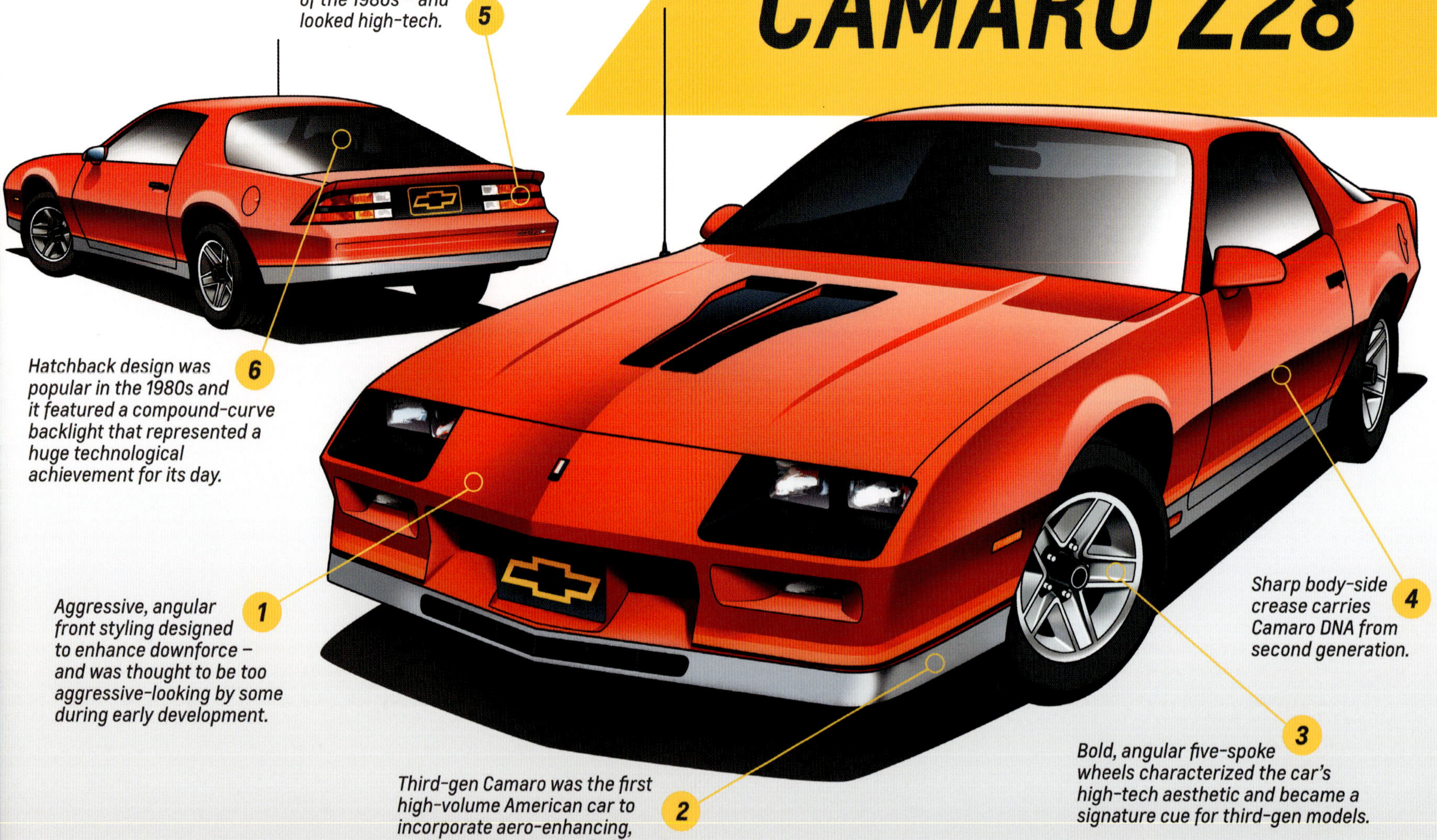

BACK TO THE FUTURE

7

Third generation: 1982–1992

While most critics couldn't say enough good things about 1970's new Camaro, not all Detroit watchers were so willing to paint a rosy picture. According to *Car and Driver*, the second-generation F-body represented a clear case of too *much*, too late. That is, Chevrolet men were blindly tossing beaucoup bucks at a market segment that was "dying on the vine." Numbers, reportedly, said it all. Pony cars, or sport compacts, had carved out 13 percent of Detroit's sales pie in 1967, but only 9.2 in 1969. That figure plummeted to 3.4 by 1972, and the vehicles themselves began dropping like overripe fruit a few years later.

"Mustang created the class in 1965, giving rise to a host of imitators," began *Motor Trend's* assessment of the situation early in 1975. "[But] in a classic misreading of market signs, each Pony Car was hastily dispatched down the road to its own demise. Mustang waxed fat and lazy, only recently shedding its husk to return to the original concept. Barracuda and Challenger elected the performance image, falling by the wayside when performance ceased to sell. Cougar ballooned into obesity and left the fold for the realm of luxury. Javelin drove in wandering circles never really finding any definition or intended direction. The only survivors are Camaro and Firebird."

So why were Chevy and Pontiac people still willing to beat *Car and Driver's* dead horse? Thoughts of having the race all to their lonesome after 1974

(opposite)
Sales of the new third-generation Camaro surged by 50 percent for 1982, the same year *Motor Trend* awarded its Car of the Year trophy to Chevrolet's new pony car. Again, a sport coupe (now fitted with a hatchback) was the only available body style. *GM*

surely enhanced interest in this segment regardless of that dire prognosis. But consider the odds stacked against them. Following 1969's artificially augmented production peak, Camaro sales drooped noticeably, to a disappointing 68,600 in 1972. While a United Auto Workers strike at the Norwood plant obviously influenced that flop, those aforementioned sagging market shares had already convinced some GM execs that it was time to put their two formerly proud ponies out to pasture.

That crippling strike began on April 7, 1972, and it was during ivory tower meetings soon afterward that Camaro's fate was supposedly sealed. "Its sales were not very high, so there was a plan going on to drop it," said Chevrolet chief engineer Alex Mair to author Gary Witzenburg eight years later. According to Camaro chief engineer Bob Dorn, Ed Cole did indeed officially announce the end of the road, in part because he didn't want to spend the required millions to meet upcoming federal crash-proof bumper standards. But the tide started to

FROM THE HORSE'S MOUTH VII

"The third-generation Camaro will always be a cultural symbol of the 1980s because its design epitomized the era's high-tech cultural trends. It also grew into more of a serious sports car and with that, its form was wrapped around a chassis designed for a new level of function. [This] Camaro represented a distinct breakaway from previous generations, which were undeniably influenced by European grand touring cars. [It featured] a uniquely American design with a form developed for function—and its aggressive front-end was deemed almost too aggressive by some in the company. Perhaps more than any other generation, the [third-gen] Camaro was a car of its time. You can see that influence in every detail, from aerodynamic exterior [touches], such as the ground effects on the Z28, to the introduction of digital instruments [inside]."

Chevrolet Global Design executive director John Cafaro, February 2015

Named Chevrolet's Global Design executive director in June 2015, John Cafaro first went to work for GM in 1977 after completing an industrial design degree at New York's Pratt Institute. He was Corvette/Camaro chief designer from 1986 to 1999. ***GM***

turn once Dorn's team proved they could perform that upgrade for a fraction of initial projections.

Also helping save the day was mucho support from various GM executives. "I was one who wanted to keep the car," continued Mair, " [But] I wasn't alone; a lot of people did. There were [those] at both Chevrolet and Pontiac who strongly wanted it, and that was important."

"Whether it was Alex Mair or John DeLorean who saved it, or the work we did, I don't know," added Dorn, "but… all of that added together to keep the car going."

Spared an untimely scrapyard visit, Camaro immediately found its way out of reverse, topping 95,000 in sales in 1973, 150,000 in 1974, 180,000 in 1976, and raring up to nearly 220,000 in 1977, the year it bested its Dearborn rival (albeit the struggling Mustang II) for the first time. On May 11, 1978, GM's Van Nuys plant rolled out Chevy's two millionth pony car, and the breed's all-time annual production high (282,571) followed in 1979.

What was it Darwin said about survival of the fittest?

"Camaro and Firebird have survived for a number of reasons," continued that 1975 *Motor Trend* report. "They are small cars in an era of demand for small cars. They are versatile enough to be operated economically when economy is mandatory. They are overpriced, but this is an age of overpricing. Their saving grace is quality."

"If we have ever achieved balance, design efficiency and beauty in a vehicle, it most certainly is in the '82 Camaro," said Camaro chief designer Jerry Palmer. "Contemporary and aerodynamic, it's in keeping with the flavor and heritage of all the great Camaros. Every surface is designed and modeled like a piece of sculpture." ***GM***

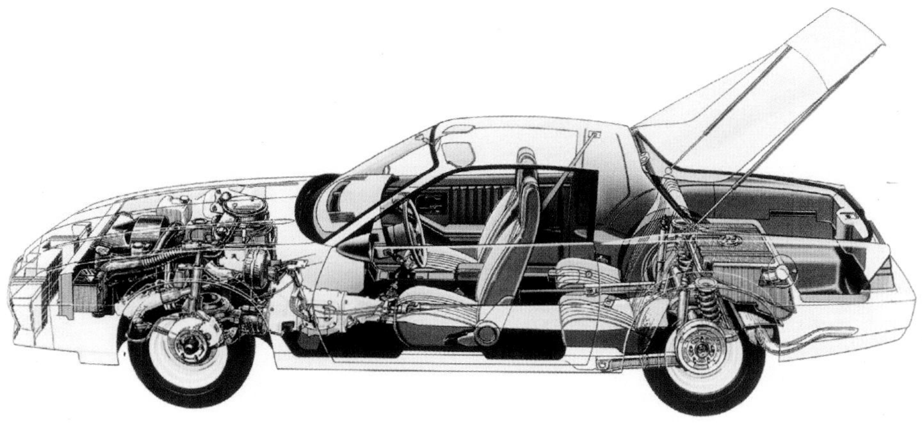

"We wanted to create a car that would reflect our technical capabilities and compete with the best the world has to offer," said Camaro Chief Engineer Tom Zimmer in 1982. "We started with the Z28, which we feel represents our handling and performance capabilities and objectives; and once having achieved those objectives, created Berlinetta and Sport Coupe." ***GM***

A Camaro first, a four-cylinder known as the "Iron Duke," was the base engine for 1982's sport coupe. The carryover Berlinetta (shown here) came standard with a V-6. A 305-cid small-block V-8 was optional in both cases. ***GM***

Saving Camaro from the same fate as Plymouth's Barracuda, Dodge's Challenger, and AMC's Javelin was among DeLorean's main goals from the moment he took over at Chevrolet in February 1969. Ever thinking outside the box, he began plotting to keep Camaro viable in an ever-changing marketplace even before completing the regeneration job Pete Estes had started. After having so little to say about the second-gen pony, he made ample noise with the third.

DeLorean's initial next-generation proposal involved combining the X-car Nova with the F-car twins on a new unit-body platform coded "K." Additional elements of his plan included shortening this foundation for use beneath a revised Vega and lengthening it for Monte Carlo and Pontiac's Grand Prix. K-car mockups first appeared in November 1969, with the Camaro version done in fastback and Kammback sport-wagon forms, both intended for 1973 debuts. K-car experiments included familiar rear-wheel- and prophetic front-wheel-drive layouts.

"The [design] could take either one so we could slowly phase into front-wheel-drive as we got rid of the conventional," said DeLorean to author Gary Witzenburg in 1980. "If they had done that, they could have saved a bloody fortune."

But neither the K-car nor its proponent stuck around Chevrolet for long, the former falling victim to Ed Cole's cold shoulder. "He had a way of destroying you without ever really saying no," continued DeLorean's recount. "He would always send things back for further study. Finally, it would get to the point where we had studied the hell out of something, and by then everyone was bored with the idea and it would just disappear. That's what happened to the K. [Cole] just further-studied it until it died."

As for DeLorean, he followed Estes up the corporate ladder in September 1972. His replacement, yet another former Pontiac man, James McDonald, ruled for only two years before also climbing those rungs, reaching GM's presidential post in February 1981. Filling in at Chevrolet in November 1974 was super salesman Robert Lund, who rode herd over actual development of a viable third-generation Camaro before moving on to Cadillac in 1982.

LEADING MAN: JERRY PALMER

Jerry Palmer always claimed he was one of the few automotive designers, domestically speaking, who actually qualified as a Detroit native. The son of a Pontiac salesman, he grew up in the Motor City doing what so many comrades in his field did during their high school days—draw cars. After graduation, he intended to head west, to attend California's prestigious Art Center College of Design, but finances simply wouldn't have it. So he stayed home and worked his way through Detroit's Society of Arts and Crafts school, now known as College of Creative Studies. As he told author Michael Lamm in 1983, machining prototype parts at a tool-and-die shop not only helped pay his tuition, but also constituted a "true education." Palmer interned at General Motors during his Arts and Crafts days then took a full-time position in a GM developmental studio after graduating in 1966. A U.S. Army tour immediately followed, as did short stints at GM subsidiaries in Japan and Europe, before his skills took him right up the ladder. In 1974, he was put in charge of the Chevrolet Three studio in Warren, Michigan, where Chevy's glass-bodied two-seater quickly became one of his main focuses. Palmer played a role in every Corvette design performed from 1974 until his retirement in 2002, including most notably the breed's major makeover for 1984. Among

Jerry Palmer was involved in every Corvette design from 1974 until his retirement in 2002. *GM*

other credits were 1982's third-gen Camaro and 1987's Beretta. These works were rewarded with a promotion to Advanced Design director in October 1986, followed later by the North American director of design position he held until he put his sketchpad away for good. He was named to the National Corvette Museum (Bowling Green, Kentucky) Hall of Fame in 2000.

In 1975, the original goals involved introducing GM's next new pony as a 1980 model. Picking up where DeLorean left off, third-gen planners initially focused on a weight-conscious front-driver, a gas-saving ideal inspired by Washington's new Corporate Average Fuel Economy standards, which mandated more miles per gallon from Detroit iron or pay the piper. Again, Bill Porter's Pontiac design team worked in conjunction with Chevrolet's Number 3 studio, headed by Jerry Palmer, to refashion their F-cars, which would share even more body structure this time.

Early scale models, completed in March 1976, featured hideaway headlights and shorter doors, the latter change made to address second-generation owners' complaints about troublesome weight and entering/exiting difficulties in crowded parking lots or cramped garages. Shortening said doors meant bringing back rear quarter glass, last seen in 1969. For the moment, that is.

Designers supposedly completed the bulk of their work by March 1977, but the final say as to which wheels, front or back, would do the driving remained unsaid, resulting in a rescheduled 1981 intro. The plot thickened further after GM design exec Bill Mitchell retired that July and Irv Rybicki stepped up in his stead. Rybicki liked what he saw in Palmer's studio save for one element: those rear-quarter windows. Retaining a single pane of side glass was a must in his opinion, hence Palmer's people had to try again, leading to another delayed debut into 1982.

Along with all the heavy-duty chassis hardware taken for granted beneath a Z28 since its 1977 revival, 1982's Z also could be fitted with the optional Cross-Fire Injection V-8, a 165-horsepower 5.0L small-block tagged LU5. *GM*

In the end, it was concluded that Camaro's heavy solid axle in back had to remain, as did V-8 availability. To compensate, and with CAFE requirements in mind, Tom Zimmer's engineering team cut away as many pounds as possible. For starters, the wheelbase was shortened 7 inches (to 101) and overall length by 10. A fully unitized body, sans that familiar front subframe, also was created, with a computer's help. The new-for-1982 unibody was, at the same time, lighter (weight savings for the entire car equaled 470 pounds) and more rigid, and it also featured a Camaro first: a rear hatchback made up almost entirely of glass. At 966 square inches, this compound-curved window was then both the largest and most complex ever installed on an American car. Concealed headlamps didn't make the final cut.

Suspension at both ends was updated, with a modified MacPherson strut design (that retained conventional coil springs) added up front. Coils (with control arms) also displaced leaf springs in back and were joined by a roll-controlling Panhard rod. Shock absorbers were no longer staggered, thanks to the addition of a single torque arm alongside the driveshaft to prevent the axle from twisting during hard acceleration. Standard power-assist for steering and brakes (front discs, rear drums) carried over from 1981, while a four-speed stick was the base transmission in all applications.

1982 Camaro	
Construction	unitized body/frame
Model availability	sport coupe w/hatchback
Price	$8,029.50 (4-cyl.), $9,665.06 (Berlinetta V-6), $10,099.25 (Z28)
Wheelbase	101 inches
Curb weight	3,195 pounds (Z28)
Suspension	Modified MacPherson strut w/coil springs &stabilizer bar, front; solid axle w/trailing arms, coil springs, track bar &Panhard rod, rear
Steering	recirculating ball w/variable-rate power assist
Wheels	14x6 (base coupe & Berlinett); 15x7 (Z28)
Tires	steel-belted radials; P195/75R-14 (base coupe); P205/70R-14 (Berlinetta); P215/65R-15 (Z28)
Brakes	power-assisted 10.5-inch front discs, 9.5-inch rear drums
Engines	90-hp 2.5L EFI 4-cylinder (base sport coupe); 102-hp 2.8L V-6 (Berlinetta); 145-hp 5.0L V-8 (Z28)
Transmission	4-speed manual; Turbo Hydra-matic automatic, optional
Axle ratio	3.42:1 w/manual trans; 2.93:1 w/automatic

Another Camaro first, a four-cylinder engine, was standard for 1982's base sport coupe. Nicknamed "Iron Duke," this Pontiac-sourced, 90-horsepower, 2.5-liter economizer featured ground-breaking throttle-body injection but was still a purely mundane mill meant only to offer cheap, durable operation. Its underwhelming performance inspired *Time* magazine in 2007 to include the Iron Duke Camaro among its 50 Worst Cars of All Time.

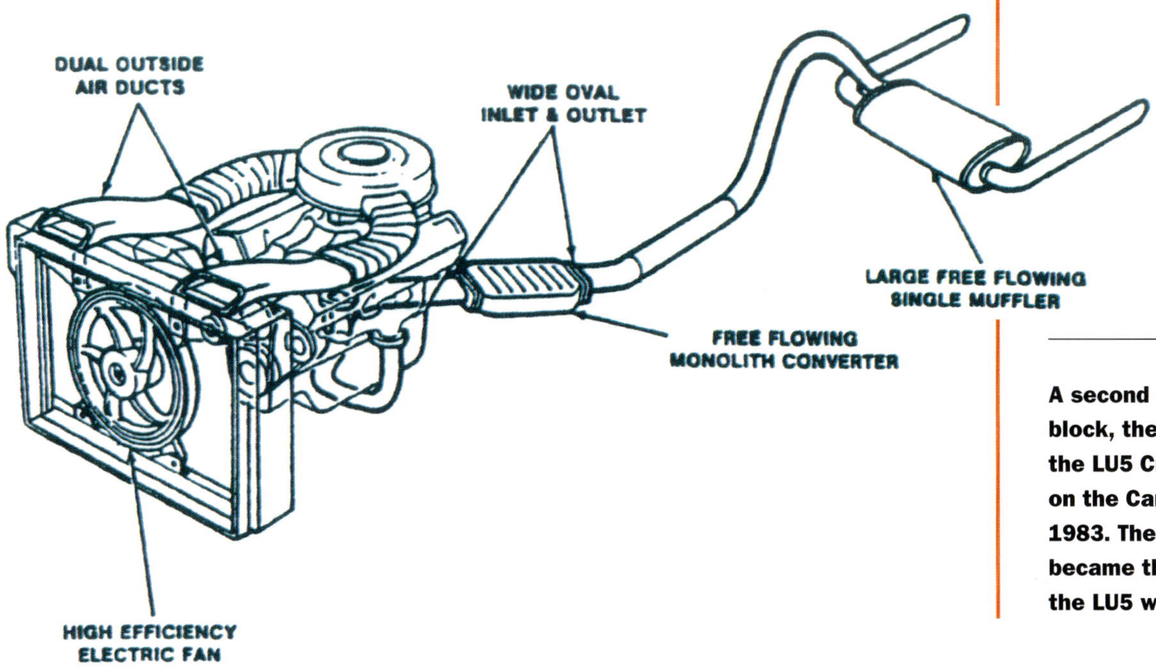

A second optional 5.0-liter small-block, the carbureted L69 HO, joined the LU5 Cross-Fire Injection V-8 on the Camaro options list in April 1983. The 190-horsepower L69 then became the top choice in 1984 after the LU5 was dropped. ***GM***

The IROC Sports Equipment package, RPO B4Z, was a $650 option in 1985. The IROC-Z became an individual model in 1988, replacing the garden-variety Z28.

"There was a time when 90 horsepower was a lot, and that time was 1932," began *Time*'s account of where this weakling ranked in automotive annals. "Fifty years later, it was bupkis, especially under the hood of Chevy's Mustang-fighter. So equipped, [1982's] Iron Duke Camaro had 0–60 miles per hour acceleration of around 20 seconds, which left Camaro owners to drum their fingers while school buses rocketed past in a blur of yellow."

Fortunately, the four-holer Camaro was dropped early in 1986. And, even more providentially, pony lovers still could put their spurs to Z28, which

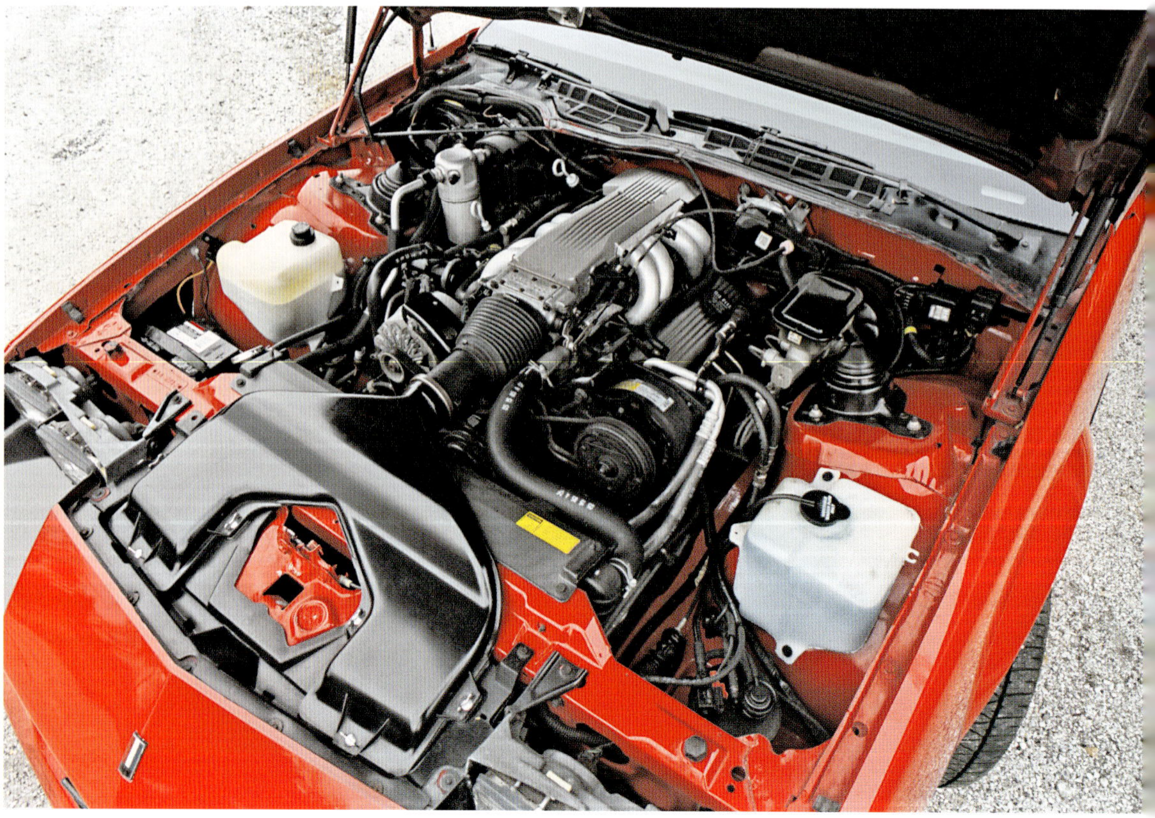

Both the carbureted L69 and Tuned Port Injection LB9 (shown here) 5.0-liter V-8s were available for 1985's IROC-Z Camaro. LB9 output was 215 horsepower.

Chevrolet ads claimed was "so new, its closest competition is its shadow." "Car of the Year" was *Motor Trend's* succinct description in 1982. "Besides being good looking, it just may be the best handling car built in America," touted *Motor Trend*'s Tony Swan.

Standard for 1982's Berlinetta coupe was a downsized 2.8-liter V-6, fed by a conventional carburetor. Entangled in CAFE concerns, Chevy's good ol' 350 V-8 didn't roll over into the third generation, leaving the 5.0-liter (305-cid) LG4 small-block to serve as the top option for the base coupe and Berlinetta. Also carbureted, the 145-horsepower LG4 was the standard heart of the third-gen Z28. Optional was the LU5 165-horsepower 5.0L fitted with "Cross-Fire Injection."

Upgraded to 175 horses, the LU5 V-8 returned as a Z28 option for 1983 then was superceded in 1984 by the L69 High Output 305, which reverted back to a conventional carburetor. HO muscle (190 horsepower) was enhanced by a twin-snorkel air cleaner and enlarged, less-restrictive exhausts featuring a wide-mouth Corvette catalytic converter last offered in 1986.

Berlinetta didn't return for 1987; in its place was a new LT option (RPO B4E) for the base sport coupe. B4E touches included sport mirrors, accent striping, custom interior, and deluxe luggage compartment. The LT deal was not available to Z28 customers. *GM*

A convertible Camaro, last seen in 1969, returned for 1987 via ASC, Inc., the former American Sunroof Corporation, founded in Los Angeles in 1965. T-top coupes (RPO CC1) were shipped to ASC, where they were converted into topless models. All 1987 Camaros—base coupe and Z28—were eligible for this treatment. *GM*

NNew for 1983 were two transmissions, the MM5 five-speed manual, and MX0 four-speed overdrive automatic. The former was standard for the Berlinetta and Z28 and optional for the base coupe, while the latter was optional for all and the only auto-trans available for Z28. A four-speed stick remained standard behind the Iron Duke alone in 1983 and 1984 before being deep-sixed in favor of the five-speed in all 1985 base installations.

Berlinetta buyers in 1984 were treated to a techno-crazed instrument panel that included digital readouts for speed and rpm, plus movable control pods on the side of those displays. "With a full-zany, 21st-century treatment, the inside is now as spacey as the outside," was *Car and Driver*'s first response to this unworldly dash. At least *Car and Driver* still considered that year's Z28 to be the best-handling car built in America.

Multiport fuel-injection (MFI) replaced the two-barrel carb atop Berlinetta's standard V-6 in 1985, but even bigger news involved the arrival of Tuned Port Injection for the 5.0-liter small-block. Tagged LB9, this TPI V-8 made 215 horsepower, in turn making this the first time a Camaro customer could take home more than 200 horses since 1974. Priced at a wallet-wilting $700, the LB9 option was offered for both Z28 and 1985's new IROC-Z Camaro but was limited to automatic transmission installations only.

Offered for 1985's Z28, the "IROC Sport Equipment Package," RPO B4Z, took its name from the International Race of Champions series, which starred

As in 1987, 1988's Camaro was available in base sport coupe (right) and Z28 forms, and the former could be enhanced with the LT package, the latter with IROC equipment, pictured at left. Sport coupes were updated with various treatments previously limited to 1987's Z28, and both models again could be converted into convertibles by ASC. *GM*

A reborn Camaro RS became a full-fledged model in 1989. As Chevrolet people explained it, this base sport coupe offered IROC-Z appearance with V-6 economy. Prices were $11,495 for the RS coupe, $16,995 for its convertible running mate. *GM*

Camaros from 1975 to 1989. Included in the B4Z deal were halogen foglamps, prominent "IROC-Z" graphics on each door, body-colored ground effects that skirted around the car's periphery, and big 16x8 five-spoke aluminum wheels wearing "Corvette-inspired" P245/50VR Goodyear Eagles.

More important were B4Z chassis tweaks. Ride height was lowered 15 millimeters (compared to a garden-variety Z28), Delco struts were re-valved for increased ride control, faster-rate jounce rubbers were added to limit wheel travel, and caster was increased from 3 degrees to 4 to compensate for those low-profile Goodyears. Gas-charged Bilstein shocks and a thicker (24-millimeter, compared to the Z28's 23) rear anti-roll bar were added, as was a recalibrated power steering system that increased steering effort to in turn improve road feel. The end result was a Camaro that handled much like a Corvette, if not better.

When the Corvette's L98 350 V-8 was repackaged in 1987 as the IROC's optional 225-horsepower B2L small-block, comparisons came even easier—both with Chevy's present two-seater and the last truly great Camaro muscle car, 1970's Corvette-powered Z28. It was time to spread the word: the breed's sporting nature was truly back.

A topless F-body, the first since 1969, returned for 1987 and was offered in both base model and Z28 forms, as well as with the IROC package. The IROC duo attained full-fledged model status the following year (replacing the "standard" Z28 in the process) and remained on the scene until Chevrolet declined to renew its contract with the International Race of Champions group in 1991. The plain-Jane Z28 returned that year and continued rolling until the F-car's 2002 cancellation.

An individual model since 1988, the IROC-Z stood in the Z28's stead up through 1990, after which time Chevrolet's contract with the International Race of Champions league ran out. Base price for the last IROC-Z coupe (shown here) was $14,555. Still converted by ASC, a 1990 IROC convertible cost $20,195. *GM*

Further enhancing those sporting ways was the 1LE Special Performance Package, created to help Camaros compete in SCCA Showroom Stock racing. This little-known option came about after GM brake engineer Phil Minch began investigating the possibility of installing bigger Caprice discs at a pony car's nose. Minch then teamed up with Camaro chief engineer Chuck Hughes and F-body powertrain manager Ray Canale to put together a complete, track-ready machine, a purposeful beast meant only for real racers. With that exclusivity in mind, Chevrolet officials made buying a 1LE Camaro a truly clandestine adventure, kinda like 1969's COPO paper chase.

Nowhere was "1LE" actually listed on RPO pages. An SCCA competitor began this seemingly classified process by requesting an IROC-Z coupe, fitted with either the B2L 350 or LB9 305, and then added RPO G92, the Performance Rear Axle option first offered for 1986's Z28. Along with G80 Positraction, the G92 deal included either 3.42:1 (manual trans) or 3.23:1 (automatic) gears.

The top-secret trick behind the 1LE purchase involved specifying the G92/IROC combo *without* air conditioning, which triggered installation of the Special Performance Package. Added now were heavy-duty four-wheel disc brakes (including those Caprice front rotors clamped by Corvette calipers); stiffer struts, springs and shocks; thicker stabilizer bars front and rear; beefed spindles and wheel bearings; an engine oil cooler; an aluminum driveshaft; and a baffled fuel tank with dual pickups to prevent engine starvation during hard cornering.

Chevrolet paperwork shows only four 1LE Camaro coupes built for 1988, with perhaps four more exported to Canada that year. Another 111 followed

1982–1992 Camaro Engines					
Year	**RPO**	**Displacement**	**Horsepower**	**Torque**	**CR**
1982	LQ9 4-cyl.[1]	151 cid (2.5L)	90 at 4,000 rpm	132 at 2,800 rpm	8.2:1
	LC1 V-6[2]	173 cid (2.8L)	102 at 4,800 rpm	142 at 2,400 rpm	8.5:1
	LG4 V-8[3]	305 cid (5.0L)	145 at 4,000 rpm	240 at 2,000 rpm	8.6:1
	LG4 V-8[4]	305 cid (5.0L)	145 at 4,000 rpm	240 at 2,000 rpm	8.6:1
	LU5 V-8[5]	305 cid (5.0L)	165 at 4,200 rpm	240 at 2,400 rpm	9.5:1
	[1]base sport coupe only (produced by Pontiac) [2]std Berlinetta, optional base sport coupe [3]single-exhaust, optional all models except Z28 [4]dual-exhaust, std w/Z28 [5]dual-exhaust & Cross-Fire Injection, optional Z28 only				
1983	LQ9 4-cyl.[6]	2.5L	92 at 4,000 rpm	134 at 2,800 rpm	8.2:1
	LC1 V-6[7]	2.8L	107 at 4,800 rpm	145 at 2,100 rpm	8.5:1
	LG4 V-8[8]	5.0L	150 at 4,000 rpm	240 at 2,400 rpm	8.6:1
	LG4 V-8[9]	5.0L	150 at 4,000 rpm	240 at 2,400 rpm	8.6:1
	LU5 V-8[10]	5.0L	175 at 4,200 rpm	250 at 2,800 rpm	9.5:1
	L69 V-8[11]	5.0L	190 at 4,800 rpm	240 at 3,200 rpm	9.5:1
	[6]base sport coup only (produced by Pontiac) [7]std Berlinetta, optional base sport coupe [8]single-exhaust, optional all models except Z28 [9]dual-exhaust, std w/Z28 [10]dual-exhaust & Cross-Fire Injection, optional Z28 only [11]HO w/Quadra-jet 4-barrel carburetor, optional Z28 only				
1984	LQ4 4-cyl.[12]	2.5L	92 at 4,200 rpm	134 at 2,800 rpm	8.2:1
	LC1 V-6[13]	2.8L	107 at 4,800 rpm	145 at 2,100 rpm	8.5:1
	LG4 V-8[14]	5.0L	150 at 4,000 rpm	240 at 4,000 rpm	8.6:1
	L69 V-8[15]	5.0L	190 at 4,800 rpm	240 at 3,200 rpm	9.5:1
	[12]base sport coupe only (produced by Pontiac) [13]std Berlinetta, optional base sport coupe [14]std Z28 [15]HO w/Quadra-jet 4-barrel carburetor, optional Z28 only				
1985	LQ9 4-cyl.[16]	2.5L	88 at 4,400 rpm	132 at 2,800 rpm	9.0:1
	LB8 V-6[17]	2.8L	135 at 5,100 rpm	165 at 3,600 rpm	8.9:1
	LG4 V-8[18]	5.0L	155 at 4,200 rpm	245 at 2,000 rpm	9.5:1
	L69 V-8[19]	5.0L	190 at 4,800 rpm	240 at 3,200 rpm	9.5:1
	LB9 V-8[20]	5.0L	215 at 4,400 rpm	275 at 3,200 rpm	9.5:1
	[16]base sport coupe only (produced by Pontiac) [17]std Berlinetta, optional base sport coupe [18]std Z28 & IROC-Z [19]HO w/Quadra-jet 4-barrel carburetor, optional IROC-Z only [20]Tuned-Port Injection, optional Z28 & IROC-Z only				
1986	LQ9 4-cyl.[21]	2.5L	88 at 4,400 rpm	130 at 2,800 rpm	9.0:1
	LB8 V-6[22]	2.8L	135 at 5,100 rpm	160 at 3,900 rpm	8.9:1
	LG4 V-8[23]	5.0L	155 at 4,200 rpm	245 at 2,000 rpm	9.5:1
	LB9 V-8[24]	5.0L	190 at 4,000 rpm	285 at 2,800 rpm	9.5:1
	L69 V-8[25]	5.0L	190 at 4,800 rpm	240 at 3,200 rpm	9.5:1
	[21]base sport coupe only, listed early but cancelled [22]std Berlinetta; optional base sport coupe (std after LQ9 cancelled) [23]std Z28 & IROC-Z [24]HO w/Quadra-jet 4-barrel carburetor, optional Z28 & IROC-Z [25]Tuned-Port Injection, optional Z28 & IROC-Z only				
1987	LB8 V-6[26]	2.8L	135 at 4,900 rpm	160 at 3,900 rpm	8.9:1
	LG4 V-8[27]	5.0L	165 at 4,400 rpm	245 at 2,800 rpm	9.3:1
	LB9 V-8[28]	5.0L	190 at 4,000 rpm	295 at 2,800 rpm	9.3:1
	B2L V-8[29]	5.7L	225 at 4,400 rpm	330 at 2,800 rpm	9.3:1
	[26]std base sport coupe & LT [27]std Z28 & IROC-Z; optional w/other models [28]optional Z28 & IROC-Z only [29]Tune-Port Injection,optional IROC-Z only				

1982–1992 Camaro Engines *(continued)*					
Year	**RPO**	**Displacement**	**Horsepower**	**Torque**	**CR**
1988	LB8 V-6[30]	2.8L	135 at 4,900 rpm	160 at 3,900 rpm	8.9:1
	L03 V-8[31]	5.0L	170 at 4,000 rpm	255 at 2,400 rpm	9.3:1
	LB9 V-8[32]	5.0L	220 at 4,400 rpm	290 at 3,200 rpm	9.3:1
	B2L V-8[33]	5.7L	230 at 4,400 rpm	330 at 3,200 rpm	9.3:1
	[30]base sport coupe only [31]std IROC-Z; optional base sport coupe [32]optional IROC-Z only [33]Tune-Port Injection,optional IROC-Z coupe only				
1989	LB8 V-6[34]	2.8L	135 at 4,900 rpm	160 at 3,900 rpm	8.9:1
	L03 V-8[35]	5.0L	170 at 4,000 rpm	255 at 2,400 rpm	9.3:1
	LB9 V-8[36]	5.0L	195 at 4,000 rpm	295 at 2,800 rpm	9.3:1
	B2L V-8[37]	5.7L	230 at 4,400 rpm	330 at 3,200 rpm	9.3:1
	[34]std RS coupe [35]std RS convertible & IROC-Z; optional RS coupe [36]optional IROC-Z only [37]Tuned-Port Injection, optional IROC-Z coupe only				
1990	LH0 V-6[38]	3.1L	140 at 4,400 rpm	180 at 3,600 rpm	8.5:1
	L03 V-8[39]	5.0L	170 at 4,000 rpm	255 at 2,500 rpm	9.3:1
	LB9[40]	5.0L	210 at 4,400 rpm	285 at 3,200 rpm	9.3:1
	LB9[41]	5.0L	230 at 4,400 rpm	300 at 3,200 rpm	9.3:1
	B2L[42]	5.7L	245 at 4,400 rpm	345 at 3,200 rpm	9.3:1
	[38]std RS coupe [39]std RS convertible; optional RS coupe [40]single-exhaust, std IROC-Z [41]dual-exhaust, optional IROC-Z coupe only [42]Tuned-Port Injection, optional IROC-Z coupe only				
1991	LH0 V-6[43]	3.1L	140 at 4,400 rpm	180 at 3,800 rpm	8.5:1
	L03V-8[44]	5.0Lr	170 at 4,000 rpm	255 at 2,400 rpm	9.3:1
	LB9 V-8[45]	5.0L	205 at 4,200 rpm	285 at 3,200 rpm	9.3:1
	LB9 V-8[46]	5.0L	230 at 4,200 rpm	300 at 3,200 rpm	9.3:1
	L89 V-8[47]	5.7L	245 at 4,400 rpm	345 at 3,200 rpm	9.3:1
	[43]std RS coupe (credit option for convertible early in year, then became standard) [44]std RS convertible early (optional later); optional RS coupe [45]single-exhaust, std Z28 [46]dual-exhaust, optional Z28 coupe only [47]Tuned-Port Injection, optional Z28 coupe only				
1992	LH0 V-6[48]	3.1 liter	140 at 4,400 rpm	180 at 3,200 rpm	8.5:1
	L03 V-8[49]	5.0 liter	170 at 4,000 rpm	255 at 2,400 rpm	9.3:1
	LB9 V-8[50]	5.0 liter	205 at 4,200 rpm	285 at 3,200 rpm	9.3:1
	LB9 V-8[51]	5.0 liter	230 at 4,400 rpm	300 at 3,200rpm	9.3:1
	B2L V-8[51]	5.7 liter	245 at 4,400 rpm	345 at 3,200 rpm	9.3:1
	[48]std RS coupe & convertible [49]optional RS coupe & convertible [50]std Z28 coupe & convertible [51]optional Z28 coupe only (B2L featured Tune-Port Injection)				

NOTE: *CR is compression ratio*

in 1989 and 62 in 1990 before the secret apparently became not-so-secret. Production jumped to 458 in 1991 and 705 in 1992. Didn't know there were that many card-carrying SCCA racers making the rounds back then.

A dual-catalyst exhaust system appeared for 1989's G92 package, priced at $466 for Clark-Kent-type air-conditioned Camaros. The price jumped to $675 when A/C was tossed aside on the way to buying a 1LE Superpony. As for the output impact of those duals, the manual-trans LB9 went from 220 horsepower to 230 with them, while the auto-only B2L jumped from 230 to 240. Now rated at 245 horsepower, all B2L 350s were fitted with dual cats in 1990.

In quieter news, Berlinetta failed to return for 1987, leaving the basic sport coupe and ever-popular Z28 to carry on. New for 1987 was the B4E LT package, which included a custom interior, deluxe storage compartment, soft-ride suspension, accent striping, and sport mirrors. A reborn RS model—combining V-6 economy with jazzy ground effects and Z28 wheels—also appeared in limited numbers for California and Florida customers. Offered again in select Southwest regions for 1988, RS became the base Camaro model the following year. A V-6 was standard for 1989's RS coupe, a 5.0-liter V-8 for its

1982–1992 Production Figures

	base spt coupe		Berlinetta	Z28 spt coupe		IROC-Z	RS	Total
1982	78,761[1]		39,744	64,882		n/a	n/a	**189,747[2]**
	[1]included 21,802 4-cylinders *[2]included 6,360 Indy Pace Car Edition Z28s*							
1983	63,806[3]		27,925	62,650		n/a	n/a	**154,381**
	[3]included 9,926 4-cylinders							
1984	127,292[4]		33,400	100,899		n/a	n/a	**261,591**
	[4]included 10,687 4-cylinders							
1985	97,966[5]		13,649	68,403		21,177[6]	n/a	**180,018**
	[5]included 3,318 4-cylinders *[6]RPO B4Z, included in Z28 total*							
1986	99,608[7]		4,479	88,132		49,585[8]	n/a	**192,219**
	[7]std 4-cyl. listed early then cancelled (none built) *[8]RPO B4Z, included in Z28 total*							
	coupe	**convt.**		**coupe**	**convt.**			
1987	83,890[9]	2639	n/a	52,863	744	38,88910	NOTE 8	**137,760**
	[9]included 794 models with optional LT treatment *[10]RPO B4Z, included in Z28 total*							
1988	66,605	1,859	n/a	n/a	n/a	27,811[11]	7,038[12]	**96,275**
	[11]included 24,050 coupes, 3,761 convertibles *[12]included in base model totals*							
	RS			**IROC-Z**				
	coupe	**convt.**		**coupe**		**convertible**		**Total**
1989	83,487	3,245	n/a	20,067		3,940		**110,739**
1990	28,750	729	n/a	4,213		1,294		**34,986**
	RS			**Z28**				
1991	79,854	5,329	n/a	12,452		3,203		**100,838**
1992	60,994	2,562	n/a	5,197		1,254		**70,007[13]**
	[13]included 8,197 w/Z03 Heritage Package							

NOTE 1: *Berlinettas were all sport coupes* ***NOTE 2:*** *1982 production total included 69,777 V-6s, 98,168 V-8s*
NOTE 3: *1983 production total included 54,332 V-6s, 90,123 V-8s* ***NOTE 4:*** *1984 production total included 98,471 V-6s, 152,433 V-8s*
NOTE 5: *1985 production total included 78,315 V-6s, 98,385 V-8s* ***NOTE 6:*** *1986 production total included 77,478 V-6s, 114,741 V-8s*
NOTE 7: *1987 production total included 60,439 V-6s, 77,321 V-8s* ***NOTE 8:*** *Included in the base sport coupe total were approximately 3,000 RS models created for sale in California and Florida* ***NOTE 9:*** *1988 production total included 42,820 V-6s, 53,455 V-8s*
NOTE 10: *1989 production total included 42,729 V-6s, 68,010 V-8s* ***NOTE 11:*** *1990 production total included 12,743 V-6s, 22,243 V-8s*
NOTE 12: *1991 production total included 31,722 V-6s, 69,116 V-8s* ***NOTE 13:*** *1992 production total included 63,556 V-6s, 6,541 V-8s*

The 1991 model lineup read RS on the bottom, Z28 on top, with that year's Z28 (foreground) receiving a new, much more dramatic rear spoiler. Body cladding for the V-6 RS (background) carried over from 1990. *GM*

convertible counterpart. Body-color ground effect panels were included in both cases.

In 1990, that V-6 was enlarged to 3.1 liters and upgraded to 140 horsepower. Production, meanwhile, decreased to the lowest annual total (34,986) to date, but for an understandable reason. A major makeover was in the works for 1991's Camaro, and Chevrolet officials just couldn't wait to raise the curtain on it.

"Its creature comforts conspicuously refined, Chevrolet's 1991 Camaro will hit the streets in March—a full seven months early," announced a January press release. "Coincident with the name change that brings 'Z28' back to prominence, Chevrolet has made the '91s quieter and smoother."

But the Bowtie boys by no means neglected the F-car's louder alter ego. Called an air-conditioned 1LE by *Car Craft*, the B4C Special Service Package Camaro debuted exclusively for 1991's RS coupe with many of the Special Performance group's goodies plus the comfort of climate-controlled operation. Others knew the B4C option as the "police package," pegged at $3,125 with the LB9 small-block, $3,950 with the B2L 350. Prices in 1992 were $3,479 with the LB9, $4,309 with the B2L. B4C production was 592 in 1991, 589 in 1992. Corvette brake calipers were added to the Special Service Package's brakes in 1992, explaining the additional addition of "1LE" identification to B4C labeling.

Rumors of a special Camaro enhanced with Corvette components (aluminum heads, a six-speed manual, etc.) to help mark the F-car's 25th birthday proved false. Chevrolet's gifts to customers in 1992 included an anniversary badge on all instrument panels and an optional Heritage Appearance Package (RPO Z03), priced at $175.

"You've got to go back 80 years to understand why [Camaro's] 25th anniversary is such a big deal," began a celebratory public relations release. "What [Billy] Durant and [Louis] Chevrolet offered to an America just emerging from the horse-and-buggy era was basic transportation with a sporting flair. What Camaro has offered latter-day motorists since its introduction in 1967 is that same unabashed spirit. Amid Chevy's vast array of mostly sensible hardware, Camaro has been the playful non-conformist, the life of the party. Ask people to name a Chevrolet car line, and more often than not they'll name Camaro first."

And the answer remains the same another quarter century later.

All 1992 Camaros featured 25th Anniversary badges inside on their instrument panels, but only Heritage Package (RPO Z03) models had a second badge on the outside, in this case on the deck lid. The red stripes seen here also were part of the $175 Z03 option. *GM*

1996
CAMARO SS

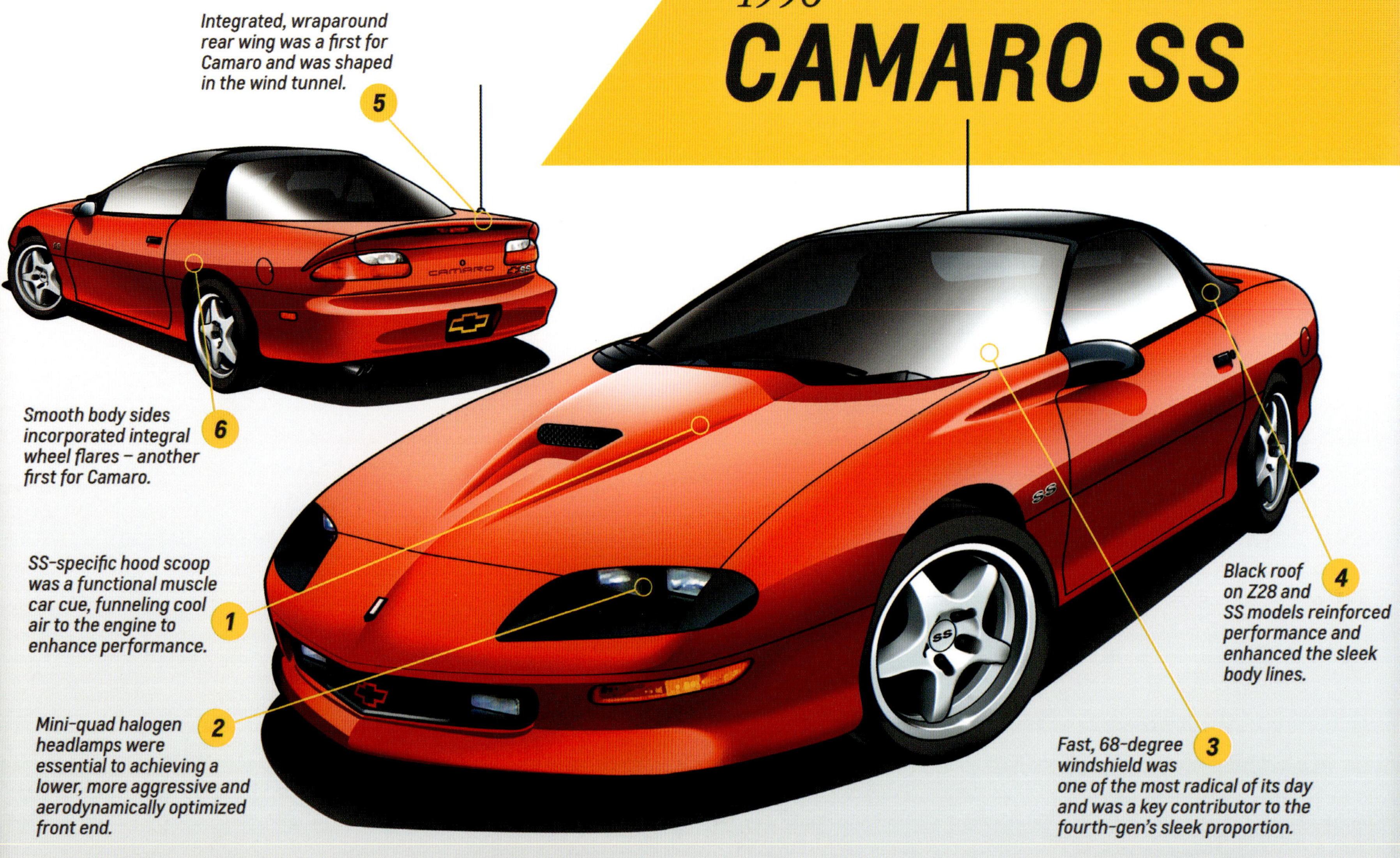

8

THE RETURN OF HORSEPOWER

Fourth generation: 1993–2002

Talk about history repeating itself. When Chevy engineers dropped their original LT-1 small-block between Z28 fenders in 1970, they helped kick start an all-new, next-generation Camaro, a way-cool sport coupe that most critics simply couldn't get enough of. They dug the way it looked, they dug the way it drove, and, in the latest Z-car's case, they dug the way it offered mucho muscle at an affordable F-body price. "Poorman's Corvette" certainly was a fair description for 1970's 360-horsepower Z28.

Fast forward nearly a quarter century to Chevrolet's renewed fourth-generation Camaro, which also wowed the F-bod faithful with its hot facade and upgraded foundation. Again, the first next-gen Z28 was fitted with a 350-cid LT1, and though its hyphen may have evaporated, the end result was fully familiar: 1993's Z once more looked like a richly endowed Corvette wearing discounted Camaro clothing.

As Chevrolet promotional people explained it, "you have to go back 23 years to the introduction of the first LT1 to understand why performance enthusiasts are so excited about the 1993 Z28. In Camaro circles, the words 'power' and 'performance' are synonymous with the 1970 LT1, and many enthusiasts have yearned for its return." Installed first in 1992's Corvette (where it made 300 horsepower), Chevy's reborn LT1 not only revived memories of muscle car days gone by, it also helped Z28 drivers in 1993 forget all about those hot ponies of yore.

(opposite)
Available as a base coupe or Z28 (both again with hatchbacks), Chevrolet's fourth-generation Camaro was built at a refurbished plant in Sainte-Therese, Quebec. An SS model, last seen in 1972, returned in 1996 with the help of SLP Engineering, based in Troy, Michigan. *GM*

1993 Z28	
Construction	unitized body/frame w/ composite body panels
Model availability	sport coupe w/hatchback
Base price	$16,779
Wheelbase	101.1 inches
Curb weight	3,373 pounds
Suspension	short/long arm (SLA) w/coil springs, gas-charged shock absorbers & tubular stabilizer, front; solid axle w/coil springs, gas-charged shock absorbers & solid stabilizer, rear
Steering	power-assisted rack-and-pinion, 14.4:1 ratio
Wheels	16x8 aluminum
Tires	Goodyear Eagle P235/55R16
Brakes	power-assisted 4-wheel disc; 10.9-inch rotors, front; 11.4-inch, rear
Engine	275-hp 5.7L LT1 V-8 w/multi-port fuel injection
Transmission	6-speed manual, std; 4-speed automatic, optional
Axle ratio	2.73:1 w/manual trans; 3.23:1 available w/automatic

"Read 'em and weep, you wanton naysayers of contemporary technology," wrote C. Van Tune after pitting an old LT-1 versus new LT1 in a *Car Craft* track trial. "The [LT1] flat kicks the tail of its beloved progenitor in every performance test, plus gets more than twice the fuel economy to boot." Zero-to-sixty went by in 5.6 clicks for the new, 7.0 for the old, while quarter-mile time slips read 14.10 seconds at 98.83 miles per hour and 14.81 at 98.04, respectively. Faster is good, right? So why live in the past when there's no future in it?

Contemporary rivals didn't stand a chance, either. Along with the LT1's 275 healthy horses—30 more than 1992's *optional* B2L 350—the fourth-gen Z28 also came standard

FROM THE HORSE'S MOUTH VIII

"More than twenty years after its debut, the fourth-generation Camaro still looks as sleek as anything on showroom floors today. It was a very aggressive design intended to evolve the proportion from the third-generation car with a provocative exterior and greater aerodynamic performance. It has a very sculptural form vocabulary that was definitely all-new for the Camaro. Having a low front end was important to the design. It really worked with the high decklid rear spoiler to enhance the appearance of motion. All these year later, it still looks contemporary—and *fast*!"

Camaro exterior design manager
Kirk Bennion, February 2015

After completing his fine arts/industrial design degree at the Cleveland Institute of Art in 1984, Kirk Bennion immediately went to work at GM as an exterior designer. He was exterior design manager for the C7 Corvette, introduced for 2014, and also was responsible for that year's Z28. *GM*

Created in 1988 at GM's Advanced Design Concept Center in Santa Barbara, California, the aptly named California Concept was powered by a V-6 engine mounted up front. Notice the mirrors, which carried over in similar fashion into fifth-gen Camaro production. *GM*

1993 Z28 V-8	
Type	OHV LT1 Gen-II small-block
Displacement	350 cubic inches (5.7 liters)
Bore	4.00 inches
Stroke	3.48 inches
Horsepower	275 at 5,000 rpm
Torque	325 at 2,400 rpm
Compression	10.5:1
Fuel delivery	multi-port fuel injection (MPFI)
Ignition	Delco-Remy Opti-Spark electronic
Cooling	reverse-flow w/5-blade electric fan
Cylinder block	cast-iron w/5 main bearings
Crankshaft	cast-nodular-iron w/4-bolt main bearing caps
Connecting rods	forged-steel
Pistons	cast-aluminum
Cam	hydraulic w/323 degrees duration on intake, 324 on exhaust
Cylinder heads	aluminum w/53.7cc combustion-chamber volume
Valve sizes (inches)	1.94-inch, intake; 150-inch, exhaust
Valve lift (inches)	0.450, intake; 0.450, exhaust
Exhaust system	cast-iron rear-dump manifolds w/single catalytic converter & muffler, dual 2.25-inch tailpipes

with a Borg-Warner T56 six-speed manual trans, four-wheel disc brakes, and big 16x8 aluminum wheels wearing fat Goodyear Eagle rubber, all wrapped up in a tidy bow stamped with an agreeable $16,779 base price. As far as performance-per-dollar was concerned, it just didn't get any better in 1993 than Chevy's new Z.

"Its specs are awesome," concluded *Motor Trend* in its annual review of the sporty segment's Best Buys in 1993. "Its price is incredible; the result, inevitable. Z28 is our Bang for the Buck winner—by the widest margin in the event's history."

Even after a total redesign for 1994, Ford's Mustang still couldn't keep up with its ever-present adversary. "It's sad that the [new fourth-gen] Mustang GT runs so far behind Z28," wrote *Car and Driver's* Kevin Smith. "Sure, [it] has taken huge strides in styling and sophistication, but at [the track] it got its haunches hammered by the archrival Camaro."

Nailing the high-perf competition, however, wasn't the new Camaro's sole claim to fame. Beneath all that performance hardware was a revamped

continued on page 164

DE-HYPHENATED

In 1990, Chevrolet's Gen-I V-8 turned thirty-five, making it rather young in people terms but certainly aged as far as automotive designs go. Sure, electronic fuel injection advances certainly had modernized the venerable small-block during the 1980s. Yet beneath all that newfangled high-tech (for then) hardware beat the heart of basically the same mouse motor that had been running long and strong since 1955. It wasn't that the Gen-I was necessarily broke, mind you; it just needed a little fixin' if it was to keep up the performance pace into another new decade.

The Corvette's 5.7-liter L98 V-8 was making 250 horses in top form by 1990, but that wasn't enough for Chevrolet's top brass. Two years earlier, they had directed engineer Anil Kulkarni to radically regenerate their thirty-something rodent, among other things instructing him to boost power by 20 percent. He didn't disappoint. Introduced for 1992, his Gen-II V-8 produced 300 horses, just as ordered. And so dramatic was this makeover that officials named Chevy's latest, greatest small-block in honor of the legacy's hottest-ever carbureted rendition, in this case sans hyphen: "LT-1" from the 1970s became "LT1" for the 1990s.

According to Corvette chief engineer Dave McLellan, the next-gen small-block earned its revered recognition because of its unprecedented muscle—it was easily stronger than the original 370-horsepower LT-1. Say what? Remember that advertised outputs prior to 1972 were gross figures recorded at the flywheel with no regard for accessory-drive drag or other external losses. Detroit's net ratings, which began showing up in 1971, were more honest numbers that reflected real power delivered directly to the road. While many of those 370 ponies never made it to the 1970 Corvette's rear wheels, all 300 did twenty-two years later, a reality demonstrated by magazine road tests. LT-1 Corvette results in 1970 ran as low as 14.17 seconds for the quarter-mile. 5.7 ticks for a 0–60 pass. Comparable timings for 1992's LT1 read 13.7 and 4.92. 'Nuff said.

As it was, comparing the LT1 with its LT-1 ancestor mattered little to drivers who were familiar only with the 250 horses propelling their 1991 Corvettes. The

Like its Corvette counterpart, Camaro's LT1 V-8 featured aluminum heads on an iron cylinder block. A third LT1, created for 1994's new Impala SS, used cast-iron heads. Output was 300 horsepower in Corvette installations, 275 in Camaro, and 260 in the Impala application. *GM*

new-for-1992 small-block truly was born again, with very little interchanging between LT1 and L98.

Block height, bore spacing, and displacement carried on per small-block tradition. Most everything else, however, was drawn up on a clean sheet of paper, beginning with the iron cylinder block and aluminum heads, which were recast to incorporate a new "reverse-flow" cooling system. Those heads also featured perfected ports that improved flow while retaining the L98's valve sizes. Helping build more muscle was a compression increase and more aggressive roller cam.

Additional Gen-II highlights included a multi-port fuel injection system (that did away with the TPI setup's exposed long-tube runners) and the innovative OptiSpark distributor, now located up front. As its name implied, this unit used optical signals to govern ignition timing. The most precise small-block ignition ever was the end result, though some bugs were encountered early on.

Various other upgrades occurred during the LT1's short run, beginning with updated cam specs on the exhaust side for 1993 that bumped maximum torque up from 330 to 340. New powdered-metal connecting rods replaced the previously used forged rods in 1994, but most notable that year was a switch to improved sequential fuel-injection system. Working in concert with a more powerful ignition, this SFI equipment enhanced throttle response and idle quality and lowered emissions but did not change advertised output.

Two variations on the 5.7-liter LT1 theme also appeared, one for 1993's Camaro, the other for Chevy's reincarnated Impala SS, introduced the following year. Rated at 260 horsepower, Impala's Gen-II V-8 relied on cast-iron cylinder heads in place of the aluminum units bolted atop Corvette/Camaro LT1s, while output in 1993 F-body applications was 275 horsepower.

A long list of mostly trivial differences set the Corvette and Camaro Gen-IIs apart, including the expected relocation of various accessory mounting points. Appearance-wise, the Camaro LT1 didn't feature "beauty shields" atop its injectors like its two-seat counterpart, and the F-car's valve covers were rather plain stamped-steel units, as opposed to the 1993 Corvette's dressier composite pieces, which had replaced the stylish magnesium valve covers used in 1992.

The most notable disparity came on the bottom end: the Camaro LT1 featured two-bolt caps for all five main bearings, while its Corvette brethren used beefier four-bolt pieces for its three center mains. Exhaust manifolds also differed, with the Corvette's traditional center-outlet layout traded for a rear-exiting design. From there back, the Camaro installation required a more restrictive exhaust system (with, among other things, one catalytic converter compared to the Corvette's dual setup), and that was the prime explanation behind the drop from 300 horses to 275.

A new cam helped push F-car Gen-II output up to 285 horses in 1996, and this figure carried over into the next year, the last for the LT1/pony car combo. Maximum muscle hit 305 horsepower for the LT1 installed in SLP Engineering's SS variants in 1996 and 1997.

Like the Impala SS, the LT1 Corvette was last seen in 1996. Offered as well that year was an enhanced Gen II, the 330-horsepower LT4, which featured new aluminum heads with taller ports and bigger valves, a meaner cam, and 10.8:1 compression. The LT4 was standard for 1996's Grand Sport, optional for other Corvettes. SLP also installed 330-horsepower Gen-IIs into 1997 SS Camaros to help commemorate the F-body's 30th anniversary.

Select Gen-II V-8s

Debut Year	RPO	CID	Bore & Stroke	Horsepower	Torque	CR
1992	LT1[1]	350	4.00 x 3.48	300 at 5,000 rpm	330 at 4,000 rpm	10.5:1
1993	LT1[2]	350	4.00 x 3.48	275 at 5,000 rpm	325 at 2,400 rpm	10.5:1
1994	LT1[3]	350	4.00 x 3.48	260 at 5,000 rpm	330 at 3,200 rpm	10.5:1
1996	LT4[4]	350	4.00 x 3.48	330 at 5,800 rpm	340 at 4,500 rpm	10.8:1

NOTE: *CID is cubic inch displacement; bore & stroke in inches; CR is compression ratio*

[1]*Corvette*
[2]*Camaro Z28*
[3]*Impala SS*
[4]*Corvette & 1997 30th Anniversary Camaro SS*

As it often does, *AutoWeek* magazine spilled the beans, almost exactly, well in advance of the fourth-gen Camaro's 1993 debut. That cover date is May, 1989. *used with permission of Crain Communications*

continued from page 161

F-platform that had expanded in nearly all directions. At 101.1 inches, the fourth-gen wheelbase remained constant, but 1993's body was longer, wider and taller than its 1992 predecessor, all the better to help make riders feel more at home.

Enhancing the feel behind the wheel was new rack-and-pinion steering (in place of the old recirculating-ball box used since 1967) and a redesigned short/long arm (SLA) front suspension that did away with the third-gen's MacPherson struts. Improved underpinnings in back again featured a traditional solid axle with two trailing arms, a track bar and a torque arm. Last but not least, all 1993 Camaros, base V-6 or LT1, came standard with driver- and passenger-side air bags (a Chevrolet first) and anti-lock brakes. Summed up, this truly was—again by a major margin—the best F-body yet, even though it wasn't the car General Motors originally wanted to build.

Various influences made the fourth-generation Camaro's drive to market into a real rollercoaster ride. At the time, GM was experiencing an economic downturn because of the recession of the early 1990s. After reporting record profits for 1988, the biggest of Detroit's Big Three posted losses two years later, and again in 1991 and 1992.

GM Board Chairman Roger Smith reached mandatory retirement age in August, 1990. Jack Smith was appointed as CEO in November 1992 and quickly went to work reorganizing an archaic bureaucracy that dated back decades to the days of William Knudsen. Smith created a leaner, meaner,

New up front for the 1993 Camaro was a traditional short/long-arm (SLA) suspension in place of the third-gen's modified MacPherson struts. Coil springs again brought up the rear. *GM*

more globally conscious GM and brought the company back to profitability—$2.5 billion in 1993 to be specific, followed by another $4.9 billion in 1994.

In addition to the economic recession that officially began in 1990, GM's financial difficulties of 1986 to 1990 were due to a 1986 U.S. tax law that eliminated car loan interest deductions, along with new national safety regulations mandating the installation of air bags or passive restraints in all cars by 1990. Consequently, by mid-1989, every automaker was feeling the pinch of these two federal actions because all car products became both more expensive to build and less affordable to Average Joe consumers and companies alike.

Coupled with ever-tightening safety standards, an equally growing need to build lighter, more efficient automobiles was making Detroit's job tougher,

Standard stuff for 1993's Camaro included air bags for passenger and driver, as well as anti-lock brakes. Corvette's 5.7-liter LT1 small-block V-8, introduced the previous year, was standard for the new Z28. *GM*

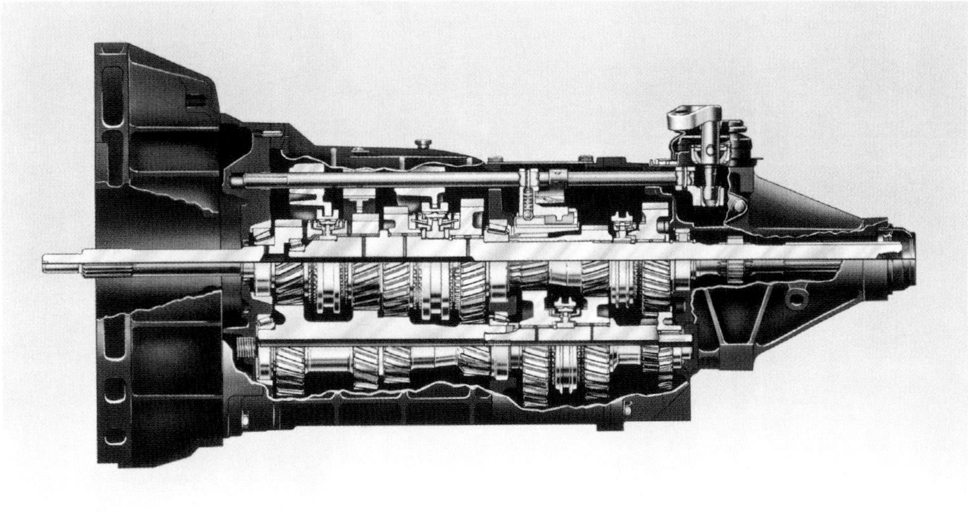

A Borg-Warner T56 six-speed manual transmission was standard for 1993's Z28. A Hydra-matic 4L60 four-speed automatic was optional. *GM*

The 3.4-liter L32—an enlarged version of GM's 3.1-liter LHO V-6—was standard for base Camaro coupes in 1993. Along with improved fuel economy, the L32 produced 20 more horsepower than its 1992 predecessor. *GM*

RS coupe and convertible models rejoined the lineup in 1996, with the idea again being to offer Z28 pizzazz at a V-6 price. The sticker for this 1997 RS droptop read $23,170, compared to a $21,770 for a base convertible and $25,520 for a topless Z28. *GM*

too, during the 1980s. Sky-high prices at the pump alone weren't fueling this new push; American firms also found themselves in a pitched battle with foreign importers, primarily from Japan, who already knew a thing or two about maximizing miles per hour, as well as selling gas-saving compacts like there was no tomorrow.

So it was that GM's initial plans for the next-gen Camaro involved a plastic-bodied, front-wheel-drive machine, something akin to the Saturn brand that went into production in July 1990. Ford folk, too, liked this idea, and even teamed up with their joint-venture Japanese partner, Mazda, to produce a FWD replacement for their aging Fox-chassis Mustang. But word got out about this "Mazstang" well ahead of its planned release, inspiring enthusiasts' complaints so vociferous that Dearborn officials opted to keep their popular pony running on until a "proper" next-gen replacement, a traditional rear-driver with an available V-8, could be delivered. In August 1987, they announced a new name for their Mazda-sourced model, Probe, which was launched in March 1988 as a 1989 model, to lukewarm responses.

A factory-direct convertible, built right on the Sainte-Therese line, appeared for 1994, the same year Ford unveiled its all-new fourth-generation Mustang. Chevy's LT1-powered Z28 (foreground) still drove circles around 1994's repackaged GT.

By contrast, Chevrolet's front-wheel-drive Camaro proposal, coded "GM80," was progressing well under the direction of chief engineer Fred Schaafsma until Ford's Probe demonstrated the marketplace's lack of interest in a front-wheel-drive performance car. Unlike most other front-wheel-drive designs of the day, the proposed GM80 actually revolved around real performance. "Schaafsma said his group had developed a front-wheel-drive car with a 325-horsepower engine," wrote John Holusha in a *New York Times* piece (September 16, 1988), headlined: "A Revival for the Rear-Wheel-Drive Car." "[That's] more power than most Corvettes have," he continued.

Unfortunately, hitching up so many hot horses to the GM80's front wheels came at a cost. Quite literally: designing front-wheel-drive architecture that could withstand that level of power using the technology that existed at the time proved prohibitively expensive. Hence, this checkbook-challenging front-wheel-drive platform was gone by 1989, replaced by a conventional rear-driven, F-body upgrade, which was initially scheduled for a 1992 debut.

"In 1989, GM reported record profits for 1988, but the forecast for 1990 was not good," reiterated former Sporty Car segment manager Harry Turner in March 2016. "By summer 1989, U.S. car sales were down industry-wide, and upper management was [again] looking for ways to reduce spending. [At the time,] the all-new 1992 Camaro was about a month behind schedule, which made it easy for management to identify it as a candidate for delay to save money. First it was a six-month delay. [But] as financial conditions worsened, it was [pushed back to] a 1993 model."

Putting a new face on the next-gen F-car was a completely contrary experience. "The 1993 Camaro was created and led by a relatively young team," said exterior design manager Kirk Bennion in March 2016. "John Cafaro, newly appointed chief designer of the Chevrolet Studio, was 32. I was 25, just out of college, and had worked for GM just two years when I began sketching the '92 (later, the '93) Camaro. We were an energetic bunch with loud rock music playing all the time. We tackled the job of working on the '91 Corvette facelift, the '93 Camaro, and what would later be the 1997 Corvette all at the same time. For myself and my co-workers, those were cherished times!"

As in 1994, the Camaro lineup for 1995 included a base coupe/convertible and Z28 coupe/convertible. Starting sticker for the V-6 coupe shown here was $14,250. *GM*

While Schaafsma's team was engaged in their losing FWD fight, Cafaro's youngsters were rockin' on with nary a worry about someone sending them back to their drawing boards. "We began sketching the '93 Camaro in 1986," continued Bennion. "One third-scale model soon followed later in the year. The sketches were distilled down to two scale models. We started a full-size clay in late 1987, with a theme from each model applied, one on the driver's side, the second on the passenger side. By early 1988 we selected a single theme. The design work then focused on addressing the production criteria that would enable the design to be mass produced. The final design was to be released to engineering in early 1989. One week after we released and photographed the final 1993 exterior, our studio photos were leaked to [the press]. The 1993 Camaro showed up on a local car magazine cover in spring of 1989. We were devastated, to say the least. This new Camaro was so wild that no one believed this was really the car."

In the end, GM reportedly invested $600 million in fourth-generation development. Was it worth it, especially considering that red-stained ledger?

"Spend more than a half-billion dollars to compete with the Mustang, which is a decade and a half old?" queried Wes Raynal in December 1992 in an *AutoWeek* article titled "Still Crazy, After All These Years." "No rational-thinking car company would do it, or even think about it. A sensible company would've long ago let Camaro slide into a well-deserved permanent sleep.

Chevrolet celebrated Camaro's 30th birthday in 1997 with the Z4C package, which featured a white exterior complemented with orange stripes. The Z4C price was $575. *GM*

[But,] having said that, the 1993 Camaro is here. Thank goodness GM [people weren't] logical enough to kill it. With all of [their] troubles, perhaps this is the perfect time for what could be considered an insane move like the new Camaro. Hail insanity."

"[Our latest] Camaro truly characterizes the Chevrolet mission, which is to provide affordable value in each segment in which [we] compete," explained general manager Jim Perkins, of perfectly sound mind, in January 1993. "[It]

LEADING MAN: JIM PERKINS

Jim Perkins wasn't the first general manager at Chevrolet to pilot an Indianapolis 500 pace car on race day; Thomas Keating beat him to the punch in May 1955. But, unlike Keating, Perkins got called back for an encore. Twice. After piloting a Beretta around the Brickyard in 1990, he toured the track again in 1993, this time at the wheel of his company's new fourth-generation Camaro. To say he loved every second of it was like calling Texas big. "There is no thrill quite like driving through the fourth turn at Indianapolis a few yards ahead of about 25,000 horsepower," he said a few months prior to his second pace-lap tour. "Nobody in his right mind would turn down the chance to do it again." Nor again; Perkins was back at Indy in 1995, leading the pack on this occasion in a Corvette pacer. A Lone Star State native (born and raised in Waco), his automotive career began at age thirteen, when he went to work at a local body shop. He bought his first car, a 1936 Chevy, the next year, and quickly became a talented hot rodder and drag racer. "By the time I was fifteen or sixteen, I was probably as good a Chevrolet mechanic as there was in Texas," proclaimed Perkins to *Automotive News*' Lindsay Chappell in October 2011. His first job with General Motors, at a Dallas parts warehouse, came in 1960. Chevrolet zone manager positions later followed in San Diego (1974) and back in Dallas (1975), as did promotions to GM dealer relations director (1977) and Chevrolet truck sales manager (1980). In 1982, Perkins moved over to Buick to

Jim Perkins drove the Indianapolis 500 pace car three times during his stint as Chevrolet general manager. He retired in May 1996. *GM*

serve as general sales manager, then left GM two years later for a group vice president position at Toyota, where he helped launch Lexus in 1989. That same year, he shocked everyone in Detroit by returning to GM as an assistant general manager at Chevrolet, followed four months later by the general manager chair. After helping revive the Bowtie brand's heartbeat, Perkins retired, at age sixty in May 1996. Still not done with the business, he moved to Charlotte, North Carolina, in 1997 to run one of America's largest retail conglomerates, Hendrick Automotive Group. Once a car guy, always a car guy.

has been redefined in such a way as to start quicker, stop quicker and handle superbly, while offering the latest in safety, environmental and anti-theft features. It even gets a little better fuel economy, despite its increased power. Camaro also is being built to extremely stringent quality standards—perhaps the most stringent in our history."

While the GM80's front propulsion didn't make the cut into reality, its corrosion-proof, dent-resistant "plastic" panel idea did. Sheet metal (two-sided galvanized steel) only appeared for the 1993 Camaro's hood and rear quarters. Made of chopped fiberglass mixed with polyester resin, sheet-molded compound (SMC) material was used to form the doors, roof, hatch, and spoiler. The compound (polyurethane reinforced with mica) front fenders and fascia were created using the Reaction Injection Molded (RIM) process. Also RIM-processed, the rear fascia consisted of Wollastokup-fortified polyurethane.

For those unfamiliar with Wollastokup, it was a trade name for a custom, silane-treated, calcium-metasilicate (actually, wollastonite) filler used to increase the durability of coatings, plastics, and rubber products. Effective March 1, 1994, the Wollastokup tag was discontinued in favor of "Wollastocoat," a registered trademark of NYCO Minerals, based in Willsboro, New York. As for wollastonite, it is a naturally occurring mineral named after British chemist/mineralogist William Hyde Wollaston, an obscure rocker who died in 1828.

Putting all these components together required an all-new factory, or at least a refurbished one. Originally opened in 1965, GM's Sainte-Therese plant

A restyled nose appeared for 1998's Camaro, available only in base V-6 and Z28 renditions as the RS didn't roll over from 1997. Buyers who wanted the new LS1 V-8 in a plain-Jane (non-Z28) wrapper could still opt for the B4C Special Service Package, which carried over from the third generation into the fourth. *GM*

At 29,009, the production count for 2001's Camaro was the lowest in the car's history. Partly to blame was an early startup of 2002 production, already recognized as the last run. For the moment. *GM*

in Quebec, Canada, was basically gutted and refitted to handle production of Chevy's bigger, better pony car, and do so with the enhanced quality control mentioned moments ago by Perkins. All this extra sweat resulted in history once again making an encore, as 1993's Z28 was chosen to pace the field for the running of the 77th Indianapolis 500, an honor earned by both the first third-gen Camaro in 1982 and the original fifteen years before.

Like its third-generation forerunner, Chevrolet's newest Camaro was initially offered only in hatchback coupe form in 1993. A fourth-gen convertible then debuted for 1994, but this time was, unlike its 1987–1992 contractor-converted predecessors, built right alongside its full-roofed siblings on the Sainte-Therese line. Nineteen-ninety-four's Z28 ragtop was introduced to the press in Kohler, Wisconsin, a curious location because…

"This is home of the renowned commode of the same name," wrote *Car and Driver's* John Phillips after Perkins's people let fly with their latest F-bod news. "Chevrolet insists this bears no particular significance, [and] we agree. Except that the lid of this car *does* raise and lower rapidly and it *is* in constant use. If you get our drift."

continued on page 176

1993–2002 Camaro Engines

Year	RPO	Displacement (liter)	Horsepower	Torque	CR
1993	L32 V-6	3.4	160 at 4,600 rpm	200 at 3,600 rpm	9.0:1
	LT1 V-8	5.7	275 at 5,000 rpm	325 at 2,400 rpm	10.5:1
1994	L32 V-6	3.4	160 at 4,600 rpm	200 at 3,600 rpm	9.0:1
	LT1 V-8	5.7	275 at 5,000 rpm	325 at 2,000 rpm	10.5:1
1995	L32 V-6	3.4	160 at 4,600 rpm	200 at 3,600 rpm	9.0:1
	L36 V-6[1]	3.8	200 at 5,200 rpm	225 at 4,000 rpm	9.4:1
	LT1 V-8	5.7	275 at 5,000 rpm	325 at 2,000 rpm	10.41
	[1]optional base coupes/convertibles & RS coupes/convertibles				
1996	L36 V-6	3.8	200 at 5,200 rpm	225 at 4,000 rpm	9.4:1
	LT1 V-8	5.7	285 at 5,000 rpm	325 at 2,400 rpm	10.4:1
	LT1 V-8[2]	5.7	305 at 5,500 rpm	325 at 2,400 rpm	10.4.1
	[2]SS				
1997	L36 V-6	3.8	200 at 5,200 rpm	225 at 4,000 rpm	9.4:1
	LT1 V-8	5.7	285 at 5,000 rpm	325 at 2,400 rpm	10.4:1
	LT1 V-8[3]	5.7	305 at 5,500 rpm	325 at 2,400 rpm	10.4:1
	LT4 V-8[4]	5.7	330 at 5,800 rpm	340 at 4,500 rpm	10.8:1
	[3]SS *[4]30th Anniversary SS*				
1998	L36 V-6	3.8	200 at 5,200 rpm	225 at 4,000 rpm	9.4:1
	LS1 V-8	5.7	305 at 5,200 rpm	335 at 4,000 rpm	10.1:1
	LS1 V-8[5]	5.7	320 at 5,200 rpm	345 at 4,000 rpm	10.1:1
	[5]SS				
1999	L36 V-6	3.8	200 at 5,200 rpm	225 at 4,000 rpm	9.4:1
	LS1 V-8	5.7	305 at 5,200 rpm	335 at 4,000 rpm	10.1:1
	LS1 V-8[6]	5.7	320 at 5,200 rpm	345 at 4,000 rpm	10.1:1
	[6]SS				
2000	L36 V-6	3.8	200 at 5,200 rpm	225 at 4,000 rpm	9.4:1
	LS1 V-8	5.7	305 at 5,200 rpm	335 at 4,000 rpm	10.1:1
	LS1 V-8[7]	5.7	320 at 5,200 rpm	345 at 4,400 rpm	10.1:1
	[7]SS				
2001	L36 V-6	3.8	200 at 5,200 rpm	225 at 4,000 rpm	9.4:1
	LS1 V-8	5.7	310 at 5, 200 rpm	340 at 4,000 rpm	10.1:1
	LS1 V-8[8]	5.7	325 at 5,200 rpm	350 at 4,000 rpm	10.1:1
	[8]SS				
2002	L36 V-6	3.8	200 at 5,200 rpm	225 at 4,000 rpm	9.4:1
	LS1 V-8	5.7	310 at 5,200 rpm	340 at 4,000 rpm	10.1:1
	LS1 V-8[9]	5.7	325 at 5,200rpm	350 at 4,000 rpm	10.1:1
	[9]SS				

NOTE: *CR is compression ratio*

STILL SEXY

The 285 horses supplied by Z28's LT1 small-block in 1996 wasn't the limit in F-body terms. Beginning that year, SLP Engineering (based in Troy, Michigan) offered its Camaro variant, Z28 SS, powered by an exclusive 305-horsepower LT1. Key to the 20 extra horses freed up by SLP was the SS Camaro's trademark hood, which featured a large functional scoop that force-fed cooler, denser outside air directly into the LT1's SFI equipment below.

An SS began life as a typical Z28 at GM's assembly plant in Sainte-Therese, Quebec. From there it was shipped complete (via GM's own carrier) to an SLP shop in nearby Montreal, where it was morphed into the muscled-up SS and sent back into GM's conventional delivery pipeline.

"In addition to offering world-class performance, another object is to make the ownership process as easy as possible," said SLP president Ed Hamburger in 1996. "For example, ordering a Z28 SS will be as simple as visiting your local Chevy dealer and asking for an order form. GMAC financing and leasing are available, and our warranty, combined with the standard coverage, provided by GM, protects the vehicle."

The SLP transformation even got its own RPO code: R7T. Along with that 305-horsepower LT1, an SS also was fitted with lightweight 17x9 cast aluminum alloy wheels wearing huge BF Goodrich Comp T/A 275/40ZR/17 tires. Convertible and T-top versions used smaller 16-inch wheels and tires and stock Z28 suspensions. SS Camaros were treated further to a larger anti-roll bar in front and a revised track bar in back. Additional suspension tweaks were optional (for coupes only), as was a free-flowing "cat-back" exhaust system and a short-throw Hurst shifter.

SLP's conversion returned for 1997 and this time also was available with the Z4C 30th Anniversary package. Both LT1 and LT4 V-8s were installed in Z4C SS models. In 1998, the SS became an "in-house" offering, listed under RPO WU8. This year's editions were creatednearly complete (now with a 320-horsepower LS1 small-block) on the Sainte-Therese line then sent to SLP, where that familiar scooped hood replaced a temporary lid installed only for shipping purposes.

Chapter 2 in the Camaro SS legacy continued until Chevrolet closed the book on its F-body in 2002. A Z4C Anniversary package again was combined with the WU8 option for that final run.

Reviving the Camaro SS was a team effort. When a Chevy customer checked off RPO R7T, a Z28 was shipped complete from the Sainte-Therese plant to a nearby SLP Engineering facility, where it was transformed into an SS. Modifications included a special hood (with functional scoop) that helped the LT1 make 305 horses instead of the stock 285. *GM*

Camaro SS Production, 1996–2002

Year	RPO	Price[1]	Total	Breakdowns
1996	R7T	$3,999	2,257	n/a
1997	R7T	$3,999	3,038[2]	included 477 6-speed coupes, 217 auto-trans coupes, 1,286 6-speed T-top coupes, 570 auto T-top coupes, 274 6-speed convertibles & 214 6-speed auto convertibles
1998	WU8	$3,500	3,017	included 2,868 for domestic distribution, 149 for Canada
1999	WU8	$3,700	4,829[3]	included 2,647 6-speed coupes, 1,380 auto-trans coupes, 365 6-speed convertibles & 437 auto convertibles
2000	WU8	$3,950	8,913	included 4,772 6-speed coupes, 2,569 auto-trans coupes, 802 6-speed convertibles & 770 auto convertibles
2001	WU8	$3,950	6,332	included 3,512 6-speed coupes, 1,956 auto-trans coupes, 411 6-speed convertibles & 453 auto convertibles
2002	WU8	$3,625	11,191[4]	included 4,811 6-speed coupes, 3,527 auto-trans coupes, 1,297 6-speed convertibles & 1,556 auto convertibles[5]

[1]For optional SS package alone

[2]Also included 957 30th Anniversary models; 108 of these featured LT4 V-8s (100 for the U.S. market, 6 for Canada & 2 prototypes)

[3]Also included 8 1LE models; 7 6-speeds plus 1 automatic

[4]Included 57 Z7D Brickyard 400 pace car convertibles (52 w/6-speed, 5 with auto trans)

[5]Also included 3,369 Z4C 35th Anniversary Editions (3,000 for domestic distribution, 369 to Canada), consisting of 1,971 coupes (1,037 w/6-speed, 934 w/auto trans), 1,398 convertibles (609 w/6-speed, 789 w/auto trans). Coupes required the CC1 T-top option.

The SS Camaro's LS1 V-8 was making 325 horsepower when the curtain came down on the fourth generation in August 2002. A 35th Edition Package (RPO Z4C, pictured here) was offered that year only for SS coupes and convertibles. The Z4C price was $2,500. *GM*

continued from page 172

Also enhancing the pot in 1993 were Chevy's 1LE Special Performance and B4C Special Service options, both rollovers from 1992. While the former was offered through 1999, the latter carried on right up until this generation's end in 2002. Again, the B4C "police package" combined the Z28's V-8 with a base coupe. Its price was $3,335 with a six-speed stick, $3,930 with a four-speed automatic. Included in the $310 1LE deal were thicker stabilizer bars front and rear, stiffer shocks, and improved cooling. Air conditioning once more wasn't available on a fourth-gen 1LE Camaro.

In other optional news, Acceleration Slip Regulation (traction control) became available, for Z28 only, in 1995, then for all models in 1999. An available 200-horsepower L36 V-6 also appeared for 1995, and this engine then became the standard powerplant for base coupes and convertibles the following year. Special anniversary packages showed up for Camaro's 30th (1997) and 35th (2002) birthdays.

In 1996, Z28's LT1 small-block was boosted up 10 horses, and a revived Rally Sport model debuted along with a new SS, supplied by SLP Engineering. The RS, converted by a division of Magna International, featured various bits of

The last fourth-gen Camaro built at Sainte-Therese, a Z28 convertible (foreground), today resides in the GM Heritage Center collection. ***GM***

LEADING MAN: SCOTT SETTLEMIRE

Scott Settlemire commonly prefers to deflect praise towards all the other great car guys who've kept Camaro running for fifty years now; from Hank Haga to Tom Peters, from Alex Maier to Al Oppenheiser, from Pete Estes to Mary Barra . . . er, make that gals, too. But where would pony riders be if not for their honored "Fbodfather?" "Scott has always had the best interests of the Camaro in his DNA," said longtime friend Harry Turner, who was segment manager for Chevrolet's F-car and Corvette from 1990 to 1995. "Camaro was more important to [him] than promotion or any personal gain. He worked tirelessly to [convey] the importance, to upper management, of the Camaro to our customers. When production ceased at Sainte-Thérèse, Scott worked [diligently] to keep management informed of the need to re-introduce a new Camaro." "[This] legacy, indeed, is very important to GM," said Settlemire in February 2016. "Camaro serves as a halo vehicle; it affects what people think about the company." Settlemire was twelve years old when his mind first focused on Chevy's pony car. It was September 10, 1966; the location was his grandmother's garage in Bakerstown, Pennsylvania. His uncle, Tom Henry, a local Chevy dealer, was hiding one of his new products, a Bolero Red Rally Sport, in preparation for its upcoming unveiling. "I was stunned by it," recalled Scott a half-century later. "It was the most beautiful car I'd ever seen." His fate was sealed then and there. After earning a business administration degree from Northwood University (Midland, Michigan) in 1975, he went to work for GM as a district sales manager (for Philadelphia and New Jersey) in September 1977. Countless other regional sales positions followed before Settlemire hit pay dirt, assistant brand manager for Camaro, in 1996. He was serving as Camaro product manager when the Quebec line shut down in 2002, then it was on to a management position for Chevrolet and Hummer Shows and Exhibits. Retirement came in November 2015, just as the sixth-gen Camaro was turning heads. Again, memorable impressions followed. "I remember seeing the car for the first time. They asked me what I thought. I said the only thing I didn't like were the air conditioning vents. I thought the [2016] was spectacular." As for his nickname, it was first heard in 1997 or 1998 at Carlisle Productions' Camaros at Carlisle event in Pennsylvania. "I was speaking with some club members from the Hudson Valley Camaro club, guys with names like Kevin Rags, Eddie da Ram and A. J. T-tops. They decided I was their 'godfather.' But one of them said, 'no, he's the *Fbod* father.'" If the concrete overshoe fits, wear it.

The apple truly doesn't fall far from the tree. The Fbodfather's mother, Elaine Settlemire, of Mars, Pennsylvania, also has been a Camaro enthusiast since the breed was born. Here she stands proudly by the fifth-gen model she's been driving since June 2009. Notice the monogram up front. *courtesy Scott Settlemire*

Scott Settlemire was twelve years old when he first fell in love with Chevrolet's pony car in September 1966. Thirty years later, he was working for GM as assistant brand manager for Camaro, a dream job if there ever was one. *courtesy Scott Settlemire*

Specialty Model Production Figures

RPO	Description	1993	1994	1995	1996	1997	1998	1999	2000	2001	2002
1LE	Special Performance Components Package	19	135	106	55	48	99	74	n/a	n/a	n/a
B4C	Special Service Package	135	668	310	228	253	239	196	254	288	708
Y87	Performance Handling Package[1]	n/a	n/a	n/a	8,734	6,918	7,291	5,304	3,365	1,697	2,412
Y3F	Sport Appearance Package	n/a	n/a	n/a	n/a	n/a	7,411	6,298	4,652	2,249	4,985
Y3B	RS Package	n/a	n/a	n/a	n/a	n/a	n/a	n/a	n/a	398	443
Z4C	Anniversary Edition[2]	n/a	n/a	n/a	n/a	4,533	n/a	n/a	n/a	n/a	3,369
Z7D	Brickyard 400 Package	n/a	n/a	n/a	n/a	n/a	n/a	n/a	n/a	n/a	57

NOTE: *See "Pace Setters," page 26, for 1993 Indy 500 Pace Car Replica numbers; See "Still Sexy," page 174, for 1996–2002 SS numbers*

[1]V-6 models only

[2]30th Anniversary in 1997, 35th in 2002

body cladding added to either a V-6 coupe or convertible. It returned in 1997, then was superseded the next year by the Y3F Sport Appearance Package, which added similar dress-ups to all models (including Z28) except SS. While the Y3F option remained through 2002, it was joined in 2001 by a new V-6 RS package (RPO Y3B), this time the work of the same SLP shop responsible for the SS. Priced at $849, 2001's Y3B option included special striping and Z28 exhausts that upped L36 output to 205 horsepower.

Camaro received a nice nose job for 1998, with hood, fenders, headlights, and front fascia all qualifying as truly new. Bosch four-wheel discs also became standard for all models this year, and electronic throttle control was added into the V-6 mix in 1999. On the V-8 side of the fence, 1998's Z28 was treated to Chevrolet's new Gen-III small-block, introduced the year before in the new C5 Corvette. In Camaro applications, this all-aluminum LS1 V-8 produced 305 horsepower, 40 less than its C5 counterpart.

1993–2002 Production Figures

	Base		Rally Sport		Z28		
	spt coupe	convt.	spt coupe	convt.	spt coupe	convt.	Total
1993	21,253	n/a	n/a	n/a	17,850	n/a	**39,103**
1994	76,531	2,328	n/a	n/a	36,008	4,932	**119,799**
1995	77,431	6,948	n/a	n/a	30,335	8,024	**122,738**
1996	31,528	2,994	8,091	905	14,906	2,938	**61,362**
1997	25,445	4,330	8,154	1,021	17,955	3,297	**60,202**
1998	31,776	2,197	n/a	n/a	17,573	2,480	**54,026**
1999	23,249	1,652	n/a	n/a	14,749	2,448	**42,098**
2000	22,097	2,945	n/a	n/a	17,383	3,036	**45,461**
2001	12,481	3,876	n/a	n/a	11,200	1,452	**29,009**
2002	14,221	2,750	n/a	n/a	19,472	5,333	**41,776**

Gen-III development dated back to late 1991. Anil Kulkarni, the man behind the LT1, was originally in charge of this project but stepped aside, giving up control to longtime small-block engineer

Ed Koerner. The Gen-III program really got cooking in 1993, with testing in early C5 prototypes beginning the following summer using mostly iron-block Gen-III "pre-runners." Most prototypes were refitted with all-aluminum counterparts by late 1995. Chevrolet then officially introduced the LS1 to the press in June 1996.

Nearly no other notable updates marked the fourth-generation's final years before the sad end came in August 2002. Chevrolet's last saleable Camaro, the second-to-last model to leave the Sainte-Therese line, was auctioned off the following September to benefit the Students Against Violence Everywhere (SAVE) charity. This fully loaded Bright Red Z28 coupe hammered out at $71,500. Meanwhile, the actual final build, a red Z28 convertible, made its way rightfully into the GM Heritage Center collection.

A sad moment, yes. But this retirement, fortunately, was only temporary. Witnessing Camaro's return to work is simply a matter of turning to page 182.

They didn't call the B4C Special Service Package the "cop option" for nothing. Shown here is what generally resulted in 2001 when Z28 power was combined with a base Camaro coupe. *GM*

TRIPLE PLAY

When the C5 Corvette appeared for 1997, it impressed critics planet-wide with its newfound comfort, convenience, and class. Performance, meanwhile, qualified as out of this world, thanks to its equally new LS1 V-8, itself a third-generation marvel that simultaneously honored tradition while leaving the past behind. Built at GM's engine plant in Romulus, Michigan, the 345-horsepower Gen-III, like all small-blocks before, was a conventional 16-valve, pushrod V-8 that also shared that time-honored "440" cylinder block layout and familiar 5.7-liter displacement label. "After all," said LS1 program manager John Juriga, "some things are sacred." Reportedly, Corvette chief engineer David Hill had considered a cutting-edge dual-overhead-cam, multi-valve engine for the C5 along the lines of the 1990–1995 ZR-1's LT5, but space constraints, as well as cost considerations, ruled out such a complicated advance.

"The LS1 has the simplicity and compactness of the pushrod layout, but with the porting so efficient and valvetrain so light and still, it breathes like an overhead-cam motor," claimed Hill. "We think a base engine at 345 net horsepower is plenty of power," added Juriga. "If that can be done with one cam, sixteen pushrods and two valves in each hole, we can live with that."

Though not necessarily state-of-the-art, the LS1 still represented a real redesign, not just a modernization. First off, compared to the Gen-II, the Gen-III's bore was less, its stroke more, changes made to allow extra cooling space between slimmer cylinders. Coolant travel was changed, too, with the Gen-II's reverse-flow system superseded by a conventional layout; Gen-III coolant once again was pumped into the block first, then to the heads. But easily the most notable break-from-the past was the LS1's all-aluminum construction, a first for a regular-production small-block. At only 107 pounds, its cylinder block weighed 53 pounds less than its iron Gen-II predecessor, and the entire engine was 88 pounds lighter.

The LS1 block at the same time was considerably stronger, thanks to extensive external stiffening ribs and its "deep-skirt" design. Unlike typical V-8 blocks that end

The new-for-1997 C5 Corvette's 5.7-liter LS1 V-8 featured aluminum heads atop an aluminum block. Additional highlights included roller lifters, a new specially tuned intake, "bat-wing" oil pan, distributorless ignition that incorporated one coil per cylinder, and double-walled, hydroformed, tubular exhaust manifolds. *David Kimble cutaway, courtesy GM*

at the crankshaft's centerline, the Gen-III unit extended below the main bearing caps, encasing its shaft in a girdle of aluminum. This extended skirt also made it possible to crossbolt the main bearing caps for additional rigidity. LS1 bearing caps featured six bolts: four in the conventional vertical location and one each running through the skirt horizontally into each side of the cap. The crank held firmly in place by those caps was cast of nodular iron and featured rolled-fillet journals for extra strength, an idea that had debuted within the Gen-II LT4 in 1996. Connecting rods were sintered forged steel, or "powdered metal," a techno trick introduced two years prior.

Cutting-edge aluminum heads went on top. Flow wizard Ron Sperry, who had developed the LT1 and LT4 heads, also was responsible for their LS1 successors, which incorporated "replicated ports." Previous small-block ports were located in two closely squeezed "Siamesed" pairs on the exhaust side, resulting in widely varying internal structures as those passages bent and turned differently with differing volumes, producing varying flow

The LS1's aluminum cylinder block (left) made its iron LT1 predecessor (right) look wimpy in comparison. Its deep-skirt design allowed those beefy main bearing caps to be crossbolted horizontally for added durability. *GM*

characteristics in the process. Keeping flow rates constant from cylinder to cylinder is vital to maximizing performance, and that's exactly what replicated ports accomplished.

"We worked hard to make sure we had all eight cylinders as close to being identical, from a geometry standpoint, as we could," explained Perry. "Each port is a continuous, runner-to-valve configuration. We get the air and fuel into the cylinder with the same level of energy from bank-to-bank and port-to-port."

Bringing up the bottom was a "bat-wing" oil pan, a shallow unit created to both offer ample capacity while ensuring a proper fit into the low-slung Corvette's engine compartment. Additional features included roller rocker arms and a roller-lifter cam. And while the LS1's sequential electronic port fuel injection was nothing new, its drive-by-wire electronic throttle control (ETC) was, at least as far as GM gasoline engines were concerned. Supplying sparks was a distributor-less ignition system featuring one coil per cylinder. Once those plugs did their job, spent gases were hauled off by hydroformed tubular exhaust manifolds.

Like the LT1 before, the Gen-III was first offered only for the Corvette, then became an F-body option (with mechanical throttle linkage) a year after its introduction. Standard for 1998's Z28 was a 305-horsepower LS1, down 40 ponies due to revised, more restrictive intake and exhaust gear (each mandated by tighter confines), and that year's Camaro SS was treated to a 320-horsepower LS1. Output for all three variants—Corvette, SS, and Z28—were bumped up five horses in 2001 to 350, 325, and 310, respectively.

Two years later, the Z06 Corvette debuted with a warmed-up Gen-III wearing yet another revered name: "LS6," used previously for 1970's 450-horsepower 454 Turbo Jet. A recast block, improved heads, 10.5:1 compression, and a lumpier cam were just a few of the improvements that transformed the LS1 into the LS6, rated at 385 horsepower. An updated composite intake manifold relied on increased plenum volume and better-flowing runners to reduce turbulence during fuel/air delivery, and this "flat-floor" intake also was added to the LS1 in 2001, resulting in those aforementioned 5 extra ponies.

By far the most powerful small-block to date, the LS6 received its own boost in 2002 to 405 horsepower. Contributing to this increase were revised catalytic converters, a free-breathing airbox, low-restriction mass-airflow sensor, and even more aggressive cam. Talk of this supreme Gen-III making its way into F-body ranks to help commemorate thirty-five years of Chevrolet pony cars proved to be just that—talk.

Select Gen-III V8s

Debut Year	RPO	CID	Bore & Stroke	Horsepower	Torque	CR
1997	LS1[1]	346 (5.7L)	3.90 x 3.62	345 at 5,600 rpm	350 at 4,400 rpm	10:1
1998	LS1[2]	346 (5.7L)	3.90 x 3.62	305 at 5,200 rpm	335 at 4,000 rpm	10.1:1
1998	LS1[3]	346 (5.7L)	3.90 x 3.62	320 at 5,500 rpm	345 at 4,400 rpm	10.1:1
2001	LS6[4]	346 (5.7L)	3.90 x 3.62	385 at 6,000 rpm	385 at 4,800 rpm	10.5:1
2002	LS6[4]	346 (5.7L)	3.90 x 3.62	405 at 6,000 rpm	400 at 4,800 rpm	10.5:1

NOTE: *CID is cubic inch displacement; bore & stroke in inches; CR is compression ratio*
[1]Corvette [2]Camaro Z28 [3]Camaro SS [4]Corvette Z06

2010 CAMARO SS

BORN AGAIN

9

Fifth generation: 2010–2015

It was late in 1996 when Camaro assistant brand manager Scott Settlemire, the fabled "Fbodfather," first got word from above that the automobile he'd loved since he was twelve was to be cut from the lineup. Initial plans involved shelving GM's F-body by 1998, but Chevrolet managed to keep the Camaro alive for another four years. Production fell by more than half for 1999.

As mentioned previously, various factors contributed to the decline in sales, including the frustrating fact that sport-utes, not sport coupes, were by then dominating the attentions of buyers young or otherwise. In addition, there were those ever-present, ever-tightening federal safety standards. The existing F-car would need an extensive redesign to meet NHTSA muster after 2002. Given steadily declining sales, it didn't seem a wise investment of corporate funds at that time.

"There was no official program to replace Camaro when the 2002 [build] came to an end," explained Settlemire in February 2016. "There had been a program called GMX280, which was Holden-based [from GM Australia], but the team felt it would not make a good Camaro. Its body lines, cowl height, and body 'tumble-home' could not be changed, and it was felt that it would make a better Monte Carlo than Camaro. It went on to become the Pontiac GTO."

(opposite)
Chevrolet's fifth-generation Camaro was offered in base LS, LT, and SS forms, all of them coupes. LS and LT models featured a 3.6-liter V-6, while the top-dog SS was equipped with a 6.2-liter V-8. ***GM***

So it was that Chevy's final fourth-generation pony car reached the end of the Sainte-Therese line, as Settlemire recalled in 2016, at 9:04 EST Tuesday morning, August 27, 2002. His memory was so precise because, fourteen years before, he had been attending a beloved uncle's funeral in Pittsburgh and was well aware of the exact moment when the dreaded call came from Canada concerning a second personal tragedy.

"It literally felt like I lost another family member," he added. "It was a very tough time, and not just for me, not just for Camaro enthusiasts, but for the people at the Sainte-Therese plant, too. It was really a bad day for all. I know cars are supposedly inanimate objects, but Camaro was more than just a car."

Countless enthusiasts agreed, as demonstrated by the printed piles of protesting emails, which reportedly reached as high as 4 feet. Always concerned with the true feelings of F-car fans, the Fbodfather even tried to assuage their grief, but in the process only drew the ire of his superiors. "I kind of got in trouble because I put the word 'hiatus' in the press release about the last car," he said. "I wanted to give enthusiasts hope that Camaro would be back; I didn't want them to lose hope."

FROM THE HORSE'S MOUTH IX

"They say absence makes the heart grow fonder, and that couldn't have been truer than as demonstrated with the enthusiasm that followed the introduction of the fifth-generation Camaro. After an eight-year absence, the return of Camaro was a thunderbolt that reignited the passion of enthusiasts around the world. It's a car for those who like to drive, and its elegant design makes you smile every time. Distilling the timeless essence of [this] design and translating into a fresh, contemporary Camaro was a challenge. The final [product] perfectly straddled that razor-sharp line between heritage and retro—clearly the fifth-generation Camaro connected with a whole new group of enthusiasts."

GM Performance Cars exterior design director Tom Peters, February 2015

Tom Peters was named GM's Performance Cars exterior design director in 2009. He was also chief designer for Cadillac's XLR roadster and Chevrolet's C6 Corvette. ***GM***

He also wasn't the only one at Chevrolet who couldn't bear to think of a world without a Bowtie-branded pony. "We did get caught working on a [new Camaro] shortly after the [F-car] went out of production but were told to immediately stop any discussion of a [next-gen model]. There [was talk] of taking the frame rails from a Corvette and marrying them to the rear 'box' of [Cadillac's] CTS, but it was a very expensive proposal and would have had limited numbers in real production."

Suffice it to say that 2003 represented a flatline segment in the Camaro timeline.

But, though the faithful didn't know it then, the legacy's y-axis began reflecting activity again in March 2004, following a meeting in England between Advanced Design Center's Bob Boniface and GM Design vice president Ed Welburn, himself a major pony fanatic. When Boniface told Welburn he wanted to begin work on a conceptual Camaro, the Global Design chief predictably said no, then softened to the point where he basically did his best Sgt. Schultz impersonation: "I see NUTHINK!" Boniface returned to this country not with

continued on page 188

All was well again when this sensational concept car debuted at Detroit's North American International Auto Show in January 2006. GM's official announcement that a Camaro comeback was in the works followed in August. *GM*

REGENERATED REDUX

Basically a bigger, better Gen-III, Chevrolet's Gen-IV V-8 debuted in LS2 form along with 2005's updated C6 Corvette, just in time to help celebrate the small-block's 50th birthday. Displacement went up to 6.0 liters—364 cubic inches for the metrically challenged—thanks to a bigger bore. And at four hundred horsepower, the LS2 stood as the Corvette's strongest base engine to date, thanks furthermore to the addition of more compression, a higher-lift cam, and increased intake and exhaust flows, by fifteen and twenty percent, respectively. It also, at the time, was the most fuel-efficient engine found in the world's top sports cars. Estimated combined city/highway economy for the new C6 was 22.6 miles per gallon.

"The Gen-IV is the best example yet of the continuous refinement in performance and efficiency that has been part of the small-block's legacy since day one," said engineering chief Sam Winegarden in October 2003. "GM has almost fifty years of experience with its valve-in-head design, and that has provided immeasurable detail for keeping the small-block a viable, relevant engine for today and the future."

An even more "relevant" Gen-IV arrived in 2006 for the C6's variation on the Z06 theme, last in action in 2004.

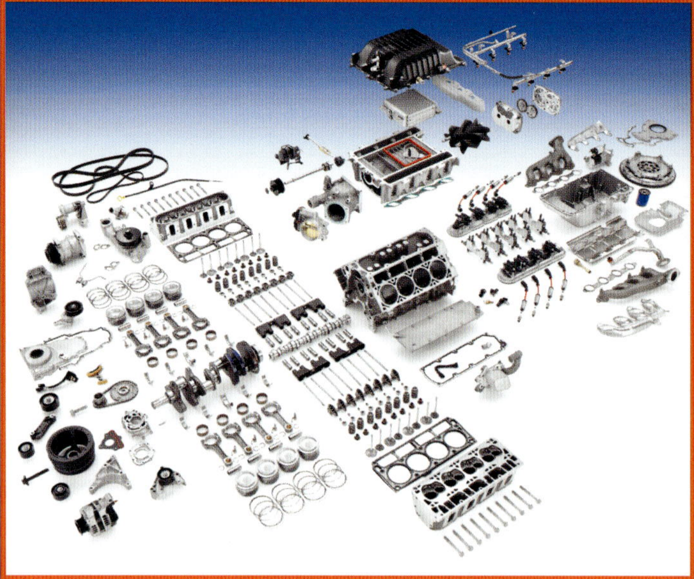

Producing 580 horsepower, the supercharged LSA small-block was introduced in 2012 as the heart of the ZL1 Camaro. *GM*

Lurking beneath the reborn Z06 hardtop's bad-to-the-bone composite shell was the LS7, an all-aluminum monster that was both the meanest small-block yet and the largest. Stretching both bore and stroke produced a familiar figure, 427 cubic inches, which translated metrically into 7.0 liters.

While its 505 max horses arrived at 6,300 rpm, this big small-block could rev confidently up to 7,100 rpm, making it one of the world's first regular-production pushrod motors able to regularly breach the seven-grand barrier and survive. The previous Gen-IV extreme was 6,600 turns for the LS2.

Chevrolet's Gen-IV small-block debuted in LS2 form for 2005's Corvette. LS2 displacement was 6.0 liters, output was four hundred horsepower. *David Kimble cutaway, courtesy GM*

"For a production engine to run at this high of an rpm blurs the lines between [pushrod] and overhead-cam design," said Dave Muscaro, assistant chief engineer for small-block engines. Blurred as well was the barrier separating road and track. "In many ways, [this] is a racing engine in a street car," continued Muscaro. "There really has been nothing else like it offered in a GM production vehicle."

LS7 heads also looked like they came right off a race track. "We consulted with our Motorsports group on numerous aspects of the design," said valvetrain engineer Jim Hicks. "The heads are simply works of mechanical art," added Muscaro. Ports were considerably larger than those in LS2 heads and were precisely massaged for exceptionally high flow.

The cylinder block was unique, too, being specially cast to preserve strength while making room for all that bore diameter. Pressed-in cylinder sleeves and tough forged-steel, crossbolted main bearing caps also set the LS7 block apart from its LS2 counterpart: the latter featured cast-in sleeves and powder-metal main caps. LS7 pistons were cast-aluminum flat-tops bolted to a balanced forged-steel crank, and lubricating all these wildly rotating parts was a competition-type dry-sump oil system, a first for street-going Corvettes.

Building the LS7 represented a job as special as the engine itself. Each was hand-assembled by a specially trained technician at GM Powertrain's new 100,000-square-foot Performance Build Center, located in Wixom, Michigan. According to this facility's site manager, Timothy Schag, this intriguing process "brings a higher level of quality, because each builder is personally involved in every aspect of the assembly."

Only about thirty LS7 V8s were produced in a day at Wixom, but quantity clearly wasn't a priority. "It was important to step away from the high-volume world we all had lived in for so long and soak in the cadence of these specialized environments," said Schag. "We established a low-volume manufacturing system on par with the world's best niche builders, but we didn't lose sight of the quality practices already in place at GM."

Another Gen-IV bore boost resulted in the 2008 Corvette's base LS3 small-block, a 6.2-liter (376-cid) V-8. High-flow LS7-style heads worked in concert with a lumpier cam, revised valvetrain (with enlarged intake valves), high-flow intake, and high-flow injectors (also copped from the Z06) to up standard C6 output to 430 horsepower. Installing the optional LS7-style dual-mode exhaust system increased output further to 436 horses.

A second hand-built 6.2-liter Gen-IV joined the LS7 at Wixom in 2009 after Chevrolet began building the highest-flying Vette yet, the 200-plus-mile-per-hour ZR1, which at $106K also represented the highest-priced GM product to date. And don't forget highest-powered. Hiding not at all beneath a transparent polycarbonate panel in the center of the ZR1's carbon-fiber hood was Chevy's LS9 small-block, a 638-horsepower bully topped by an Eaton R2300 twin-rotor supercharger specially designed to fit snugly beneath the Corvette's hood.

A year later, the LS3 went to work, along with its L99 derivative, powering SS versions of the reborn Camaro. And in 2012, the LS7 was repackaged in the new Z28.

Select Gen-IV V-8s

Debut Year	RPO	CID	Bore & Stroke	Horsepower	Torque	CR
2005[1]	LS2	364 (6.0L)	4.00 x 3.62	400 at 6,000 rpm	400 at 4,400 rpm	10.9:1
2006[2]	LS7	427 (7.0L)	4.125 x 4.00	505 at 6,300 rpm	470 at 4,800 rpm	11:1
2008[3]	LS3	376 (6.2L)	4.06 x 3.62	430 at 5,900 rpm	424 at 4,600 rpm	10.7:1
2009[4]	LS9[7]	376 (6.2L)	4.06 x 3.62	638 at 6,500 rpm	604 at 3,800 rpm	9:1
2010[5]	L99[8]	376 (6.2L)	4.06 x 3.62	400 at 5,900 rpm	410 at 4,300 rpm	10.7:1
2010[5]	LS3[9]	376 (6.2L)	4.06 x 3.62	426 at 5,900 rpm	420 at 4,600 rpm	10.7:1
2012[6]	LSA[7]	376 (6.2L)	4.06 x 3.62	580 at 6,100 rpm	556 at 3,800 rpm	9:1
2014	LS7[10]	427 (7.0L)	4.125 x 4.00	505 at 6,300 rpm	470 at 4,800 rpm	11:1

NOTE: *CID is cubic inch displacement; bore & stroke in inches; CR is compression ratio*
[1]Corvette & SSR sport truck [2]Corvette Z06 [3]Corvette
[4]Corvette ZR1 [5]Camaro SS [6]Camaro ZL1
[7]Supercharged [8]Automatic transmission [9]Manual transmission
[10]Camaro Z28

GM unveiled a topless Camaro concept vehicle in January 2007. *GM*

continued from page 185

explicit approval, mind you, but with the implication that his boss couldn't hit the kill switch on something he didn't know about.

Secret work on Boniface's concept, to be based on Holden's Zeta foundation, then quickly commenced in GM's Advanced studio, with a scale model appearing in October. But, as continuing bad luck would have it, consideration for Zeta development on the North American market evaporated about the same time, due to General Motors' percolating economic woes, soon to boil over yet again in historic fashion. These setbacks, however, were only temporary—both for the corporation's bank book and Camaro's bloodline.

Boniface's model came back into focus after Advanced Engineering director Mike Abelson began reinvestigating Zeta applications on this side of the planet. A full-size clay was ready for review by April 2005, just as GM leaders—including Welburn and GM vice chairman Bob Lutz—were meeting to consider concept vehicle proposals for 2006. A real crowd-pleaser, one that promoted the Bowtie heritage big-time, was discussed, but wearing what badge? No longer pretending to support any anti-pony stance, Welburn concluded that there was only one viable choice: Camaro. Design season then officially opened.

Although it represented a great start, Welburn immediately concluded he wasn't completely satisfied with the clandestine work of Boniface's team. So he set up a second group headed by Tom Peters in Bill Mitchell's old Studio

LEADING MAN: ED WELBURN

Once upon a time, American automakers all turned to aftermarket companies—"coach builders"—for their bodies, focusing their efforts primarily on engineering aspects. But then, in June 1927, General Motors president Alfred Sloan hired California-customizer-to-the-stars Harley Earl to head his newly opened Art &Colour section, with the idea being to cut out the middle man, to fashion all cars in-house. Already a legend, at least on the West Coast, Earl quickly put GM products at the forefront as far as factory-direct looks were concerned. "Styling" became the department name in 1937, and Earl was elevated to corporate vice president in 1940. His group's shingle became "Design Staff" in 1956, two years before he stepped down in favor of his hand-picked successor, William Mitchell. Design VP until his retirement in 1977, Mitchell was followed by Irv Rybicki (1977–1986), Charles Jordan (1986-1992), and Wayne Cherry (1992–2003). Next came Ed Welburn, the first African-American to rise to such lofty heights in Detroit. Born the son of a body shop owner in the Philadelphia area in December 1950, Welburn was eleven when he wrote to GM asking about how he could build a future as an automotive designer. An encouraging return letter gave him initial direction, and in 1972 he completed a fine arts degree at Howard University in Washington, D.C. He served an internship at GM while still at Howard then became an associate designer at the corporation's Advanced Design Studio after graduation. A position with Oldsmobile followed in 1973, and in 1989 Welburn was named chief designer at Olds. A two-year stint at Saturn followed in 1996, taking him to Germany, where he gained experience with ideas that worked planet-wide. Next came the Advanced Design director's position back in Warren, Michigan, in 1998, and the executive directorship of design, body-on-frame architecture, in January 2002. On October 1, 2003, Welburn took the baton from Wayne Cherry, this time as vice president of GM Design North America. But there remained another step. On March 1, 2005, he was named the corporation's first vice president of Global Design, putting him in charge of ten studios located in seven countries around the world—including the one in Warren where the Camaro concept was just then taking shape. The timing couldn't have been more perfect, both for the pony car awaiting new life and the design executive who already had a finger on this breed's pulse. According to Advanced Design's Bob Boniface, Ed Welburn was the fifth-generation's "patron saint." Not to mention its proud papa.

Ed Welburn became the sixth head of GM Design in 2003, following Wayne Cherry (1992–2003), Charles Jordan (1986–1992), Irv Rybicki (1977–1986), Bill Mitchell (1959–1977), and Harley Earl (1927–1958). *GM*

X, located in a Tech Center basement in Warren, Michigan. Under Peters's direction, Steve Kim, Vlad Kapitonov, and Sangyup Lee put their heads together to draw up just the startling look Welburn wanted—respectfully reminiscent of the first-gen 1969 model.

Everyone involved with the project recognized from the get-go that following in Ford's hoofprints—that is, pouring on nostalgia with a ladle akin to 2005's Mustang makeover—wasn't the answer. "You wouldn't want to repaint the Mona Lisa," said Settlemire. "The beauty of the concept was that it captured the spirit of the design, but it was a modern car." Furthermore, again in his words, "it was simply breathtaking."

"That fact that Camaro has been out of production for a number of years made it particularly important that the Camaro concept honors [that] heritage in the right way," said Boniface in January 2006. "The concept is designed to have that same broad appeal, with unmistakable style, spirit and performance," added Welburn. Last but certainly not least, the concept also spoke for itself. According to Peters, "the overall proportions, long hood and powerful fender forms say, 'this is a front-engine, rear-wheel-drive performance vehicle.'"

And the proud "'Father". . . ? "Personally, I felt the car was a winner from the time I saw it first in clay. It nearly brought me to my knees. But my feelings didn't matter. The public's *did*." And no way were they going to take sides against the family, nor overlook an offer that simply couldn't be refused.

Vito Q. Public and friends got their chance to kneel in reverence on January 9, 2006, at Detroit's North American International Auto Show. Chevrolet built two Camaro concepts (in short order) for this year's show circuit; a fully functional

Credit for sketching the Camaro concept went to South Korean-born Sangyup Lee. According to his boss, Tom Peters, Lee's work reminded him of a "street-fighting dog." And that was a good thing. *GM*

vehicle, done in silver, and a crimson-painted mockup. While the non-running Precision Red Camaro was making its static debut at the Los Angeles Auto Show, its living, breathing sibling was hitting the Motown stage with a 400-horsepower LS2 V-8, a six-speed stick, and fully independent suspension. Raves ensued.

Tom Peters (lower left) oversees Sangyup Lee's work on the Camaro concept clay model. *GM*

"The reaction of the crowd on the reveal of the concept car proved that great design is emotional and can bring out strong feelings within the community," said Settlemire. But the really big response came from *AutoWeek*, which extended its Best in Show award to Chevy's reborn Camaro, along with runner-up prizes in the Best Concept and Most Significant vehicle categories.

"Only twice in *AutoWeek's* long and illustrious tradition of awarding favorite auto show models with shiny trophies has a car won by unanimous vote. The first? That would be the Porsche Boxster concept back in 1993, which puts the Chevrolet in pretty darn good company," Settlemire said. "We certainly aren't alone in our admiration. At one point during the press review, it looked as though the number of top-tier designers from competing automakers out-numbered members of the media swarming about the car. That had to have the Bowtie brass smiling. And if the Corvette LS2 V-8 found under the concept's hood

GM Design head Ed Welburn was more than willing to show off his 1969 Camaro (center) in comparison to two concept proposals in 2005. *GM*

Appearances in director Michael Bay's various *Transformers* movies didn't hurt the Camaro cause in the least. A customized production model co-starred in the 2009 sequel, *Revenge of the Fallen*, after a modified Holden Monaro (shod in Camaro concept body panels) played the "Bumblebee" role in Bay's original film, released during the summer of 2007. *GM*

makes its way into a production rendition, we will be smiling right along with them. Of course, we have to wait a few years to see if it really happens. Fingers crossed."

In common show car fashion, no promises came standard with this one, either. "Until the concept was revealed to the public in January 2006, there was no real tangible program to bring Camaro back," said Settlemire. Among the earliest "intangible" developments involved those mates Down Under at Holden, who began exploring Zeta applications for this program (coded GMX521) about the same time their counterparts in Warren were assembling their silver concept using Sigma platform components copped from Cadillac parts bins. As for corporeal moments here in the Western hemisphere, ivory tower approval of the vehicle program initiative (VPI), a vital initial stepping stone, came in July 2006.

Crude test mules were up and running soon afterward, followed by construction of Integral Vehicle Engineering Release cars in September 2007. About 150 IVER prototypes were built to put production practices under the proverbial microscope, and another 50 partial builds were performed for crash tests and such.

Lead Designer Peters and Chief Engineer Al Oppenheiser, along with an incredibly talented team of executives, engineers,

A 426-horsepower 6.2-liter LS3 small-block V-8 (shown here) was included with manual-trans SS models in 2010. Automatic SS Camaros were fitted with a 400-horsepower L99 V-8. *David Kimble cutaway, courtesy GM*

and product managers, pulled out all the stops to create the same magic as the team that introduced that first Camaro for the 1967 model year. They would create another category killer. "Suffice it to say, GM has put its finest on this most important vehicle," said Settlemire, in August 2006.

A market test of sorts also occurred during the summer of 2007, courtesy of movie director Michael Bay, Paramount Pictures, and Steven Spielberg's DreamWorks studio. Released on July 3, the first of Bay's four box-office-busting *Transformers* films featured "Bumblebee," a clapped-out, mostly yellow 1977 Camaro that transformed (get it?) into a gleaming, glorious super car to help battle the evil Megatron and his Decepticons over the fate of the world. Playing Bumblebee's alter ego was a new Camaro… err, a Holden Monaro chassis draped in a black-striped yellow body formed from the same molds that created 2006's award-winning concept.

A customized production model played Bumblebee in 2009's sequel, *Transformers: Revenge of the Fallen*, followed by an amber-tinted SS in 2011's

continued on page 197

Standard beauty beneath the 2010 Camaro's skin included variable-ratio rack-and-pinion steering, four-wheel anti-lock disc brakes, and 4.5-link independent rear suspension. *David Kimble cutaway, courtesy GM*

WHERE THE STREETS HAVE NO NAME

Consider the plight of most birds. In the air, they're as graceful as all get-out, able to go anywhere their miniscule machine-gunning hearts desire—up, down or otherwise—at at any given second. But not so much on the ground, where the bulk of Mother Nature's breeds hop around on backward knees like a child's wind-up toy gone nuts. Fortunately, all worms, grubs, and other small Earth-bound edibles are even less agile down in the dirt.

Much the same can be said (about that split personality, not dining habits) for man's metallic winged wonders, which do their best work off this planet, not on. Simply stated, compromise is the case in all aircraft designs, as traveling higher and faster always translates into trouble at slow speeds and low altitudes.

Consider Lockheed's U-2 spy plane, the "Dragon Lady" in U.S. Air Force lexicon. Talk about your Doctor Jekyll/Mr. Hyde scenario. On one hand, this venerable recon-master literally jumps off a runway, soars nearly into space, and stays on duty for hours upon hours while piling up miles by the thousands—exactly as legendary designer Kelly Johnson intended sixty-two years ago. Getting back to the barn, however . . . well, that's something the U-2 just wasn't built to do, at least not with any measure of grace. Nor ease.

More or less a majorly modified F-104 Starfighter with its stubby "wings" ("fins" were more like it) traded for a nation-wide 103-foot lateral span, the U-2 originally wasn't even meant to include landing gear, a proposed compromise that saved weight, and in turn helped ensure a flight ceiling of more than 70,000 feet. No 1950s interceptor from any country could come close to that performance envelope, and supposedly Russian radar couldn't reach that high, either. As for getting up there and back, Lockheed's initial design included a detaching wheeled dolly for takeoffs and a belly-mounted skid-plate for landings.

The CIA six decades back also figured Soviet missiles couldn't attain such heights, a faulty fact disproved the hard way by pilot Gary Powers. At the time flying for the Central Intelligence Agency, Powers and his U-2 were blown out of the sky over the USSR in May 1960, leading President Eisenhower's people to abandon their ongoing

Various high-performance cars have served as U-2 chase vehicles over the years. Here, a fourth-generation B4C Camaro handles this duty with aplomb. "We knew they were using Camaros as U-2 chase cars," said Scott Settlemire in February 2016. "Needless to say, we were very proud about this. I wanted to mention the U-2 chase cars [in our brochures], and I even proposed letting the U.S. government lease them for $1 a year. But they never followed up on this offer." *U.S.A.F. photo, by Staff Sgt. Justin Jacobs*

cover story that U-2 flyovers involved nothing more than weather research. Powers somehow survived the strike, as well as the succeeding fall from twice the height of a typical commercial flight, and was exchanged two years later for an ensnared Russian spy.

Like all U-2s before and since, Powers's high-altitude flying camera did, of course, feature landing wheels. But that inclusion constituted a contrary compromise, one that contributed big-time to this odd duck's hesitance to find its way back into its hangar.

No way could those spindly wings house any touchdown gear, so the only choice was to use a bicycle layout, one wheel in front of the other, in a line down the fuselage's longitudinal axis. While the much larger front wheels are fixed along that axis, the rears are linked to the rudder to allow steering while taxiing. Balance during takeoff is provided by a pair of "pogo" wheels mounted in mid-wing sockets. Once up, up, and away, these auxiliary rollers fall back to base, hopefully to be reused. Handling their job during landing are a pair of titanium skids, one eneath each wingtip, to protect whichever tip strikes the unway as braking takes over—hence that "Dragon Lady" ickname. When the dust clears, ground crew then reinstall wo pogos, allowing a U-2 to taxi wherever it likes without ny draggin' whatsoever.

A rocket scientist isn't needed to point out the flaws ere. As daredevil Evel Knievel once demonstrated, broken one by bone, landing a bike from high places is never easy. And in the U-2's case, that "rear-steer" aspect makes the ask even more unwieldy. Touching down on those tiny vheels in back is required to guarantee a straight path lown a runway's center, but how's a pilot to know he/she as sufficient pitch-up attitude as the ground comes up to neet them?

Compounding the situation are super-heavy controls nd the helmeted space suit worn by U-2 crew. Not needed n the negligible atmosphere at 70K, hydraulic assist does ot come standard, meaning a real workout awaits at lenser lower altitudes. But definitely required during higher-han-the-sky flights, that fully pressurized cocoon, working n concert with an already cramped cockpit, severely mpedes a pilot's vision. Suffice it to say that looking over a houlder ain't gonna happen.

As it is, U-2 drivers already have enough on their plates vithout worrying about which truck touches down first or whether or not even a slight roll is putting a titanium tip in harm's way. In yet another compromise, all that lift provided by that humongous wingspan means a Dragon Lady simply refuses to come all the way down to Earth under power. An unrelenting ground-effected cushion of air remains in the way until the wings go into full stall, and that won't occur until all thrust goes away first. Like a navy prop-jock targeting a carrier deck in the Pacific seventy-plus years back, a U-2 pilot simply shuts down the throttle when the runway nears, dropping like a sparrow nailed through the eye by a Red Rider—toting youngster.

"We say the U-2 is the easiest plane to fly between six inches and sixty thousand feet" explained USAF spokesman Drew Buchanan to the *Los Angeles Times* in November 2012. "But it's the most difficult plane to land." So difficult, the air force long ago instituted a back-up plan, which brings us to why so many words have been expended here about a military aircraft in a book about Camaro.

From the beginning, U-2 landings have been accompanied by chase cars, with the goal being to supply ground-based direction via mobile radio. Manned by U-2 pilots themselves, these vehicles stand at the ready as a Lockheed spy plane approaches, then hot-foot it down the runway in pursuit, radioing altitude and attitude info along the way until all is down and stopped, preferably safely.

Fifth-gen Camaros picked right up where their forerunners left off. "The U-2 is the only plane in the world that has a co-pilot in a car on the ground," added Scott Settlemire. *USAF photo, by Airman 1st Class Bobby Cummings*

WHERE THE STREETS HAVE NO NAME *(continued)*

"[This] is the only aircraft in the Air Force that has a [person in] a car driving behind it talking a pilot through a landing," said Maj. Luke Lokowich, of the 1st Reconnaissance Squadron. Or, as Camaro guru Scott Settlemire explained, "it's the only plane in the world where a copilot is in a car."

Reportedly, pickups and Ford station wagons were used first back in the 1950s, followed later by El Caminos, which remained in service until 1985. Mustang cop cars came next, followed by fourth-generation Camaros equipped with Chevrolet's B4C police package. Pontiac G8s and GTOs were inducted onto the chase team following the turn of the millennium, then the 426-horsepower Camaro was signed up not long after the fifth-generation's debut.

What do U-2 crewman feel about this practice?

"[You come back] from a long mission flying for nine or ten hours, you are going to be tired [and] stagnant, and your visibility is really hindered in the suit," explained instructor Maj. Peter Van Pelt. "Your ability to feel [and] your dexterity [are diminished]. Couple that with the fact this plane is really difficult to land on a good day [and] it's really nice to have an extra set of eyes outside the airplane."

"This job is awesome," added Major Lokowich in reference to his ground duty. "You get to [drive] 110 miles per hour every day." And with no worries about citations—after all, how can a constable write a ticket on a street without a name?

Fourteen USAF U-2 instructor pilots took time off to pose along with a two-seat U-2s in August 2002 at Beale Air Force Base in California. USAF people like to point out that Super Bowl ring-wearers outnumber those who have piloted the Dragon Lady during its sixty-year-plus flight career. *USAF photo, by John Schwab*

continued from page 193

Dark of the Moon and a majorly modified Camaro in 2014's *Age of Extinction.* We can only imagine what awaits movie-goers when Bay's fifth *Transformers* epic hits theaters in June 2017 to take GM product placement practices to all new heights.

"Being a part of the *Transformers* franchise is an incredible way to showcase the design work of which GM is capable," said Ed Welburn in June 2014. "This global series gets our cutting-edge designs in front of more potential customers than we could through traditional methods."

"Placing the car as a star in the *Transformers* movies was key to getting [it] out to the public so people could have an awareness of the new Camaro," added Settlemire in February 2016. "Every child who saw the early production cars was convinced they were looking at 'Bumblebee!'" How many of those youngsters will grow up into devoted pony riders? Again, we can only imagine.

GM chief Rick Wagoner officially announced a comeback for Chevy's pony car on August, 10, 2006. "As evidence that we're not completely brain-dead, GM will build the Camaro," he said with a grin. Public introductions came at Warren's Tech Center and Southern California's Advanced Design studio on July 21,

"We intend to make the all-new Camaro relevant to younger buyer while retaining its appeal to current fans," said Chevrolet general manager Ed Peper in August 2006 while announcing that Chevy's beloved pony car indeed would return. Catching Mustang in the sporty car segment sales race surely proved the Bowtie gang did meet this goal and then some. *GM*

While some nostalgic cues did show up outside, there was next to nothing old-fashioned about the fifth-generation Camaro interior—save, perhaps, for instrumentation styles. *GM*

2010 Camaro	
Construction	unitized body/frame
Model availability	coupe
Price	$23,530, LS; $24,730, 1LT; $27,430, 2LT; #31,795 1SS; $34,500, 2SS
Wheelbase	112.3 inches
Curb weight (pounds)	3,750, LT/auto; 3,741, LT/manual; 3,769, LS/auto; 3,780, LS/manual; 3,913, SS/auto; 3,860, SS/manual
Suspension	double-ball-joint, multi-link struts w/progressive-rate coil springs & direct-acting stabilizer, front; 4.5-link independent w/progressive-rate coil-over shock absorbers
Steering	variable-rate rack & pinion
Wheels (inches)	18x7.5, steel (LS); 18x7.5 in front, 19x8 in back (LT, aluminum); 20x8 in front, 20x9 in back (SS, aluminum)
Tires	P245/55R18, all season (LS); P245/55R-18 in front, P245/50R-19 in back (LT, all season); P245/45R-20 in front, P275/40R-20 in back (SS, summer)
Brakes	Four-wheel discs w/ABS & ventilated front rotors; single-piston calipers, front & rear (LS, LT); 4-piston Brembo aluminum calipers, front & rear (SS)
Brake sizes	12.64-inch in front, 12.4-inch in back (LS, LT); 14-inch in front; 14.4-inch in back (SS)
Engine	304-hp LLT V-6 (LS, LT), 400-hp L99 V-8 (SS w/auto-trans); 426-hp LS3 V-8 (SS w/manual trans)
Transmission	Aisin Warner AY6 6-speed manual, std w/LS & LT; Tremec 6-speed manual, std w/SS; 6L50 Hydra-Matic automatic, optional (LS, LT) ; 6L80 Hydra-Matic automatic, optional (SS)
Axle ratio	3.27:1, std (LS, LT, SS/auto); 3.45:1 (SS/manual)

2008, production at GM's Oshawa plant started on March 16, 2009, and public sales commenced the following month. Chevrolet general manager Ed Peper explained further (back in August 2006) that the 2010 Camaro would target each sex, as well as both those who had first fallen in love in Detroit four years prior and long-longing enthusiasts who still remembered F-cars of yore.

"We intend to make the all-new Camaro relevant to younger buyers while retaining its appeal to current fans," he said. "The beauty of the best Camaros [of the past] is that they have always been beautiful to look at with performance that rivals expensive European GTs. Yet they were practical enough to drive every day and priced within the reach of many new-car buyers."

GM people also weren't about to forget what Chevy's show-stopper meant to Mr. and Ms. Public in 2006. "The production car was so very close to the concept, and I think that was key," explained Settlemire. "Bob Lutz was very clear in direction—do not change the car any more than is absolutely necessary for production."

2010 Camaro Engines		
Type	LLT DOHC V-6	L99/LS3 OHV Gen-IV V-8
Displacement	3.6 liter (217 cubic inches)	6.2 liter (376 cubic inches)
Bore	3.70 inches	4.06 inches
Stroke	3.37 inches	3.62 inches
Horsepower	304 at 6,400 rpm	400 at 5,900 rpm (L99); 426 at 5,900 rpm (LS3)
Torque	273 at 5,200 rpm	410 at 4,300 rpm (L99); 420 at 4,600 rpm (LS3)
Compression	11.3:1	10.7:1
Fuel delivery	direct high-pressure fuel injection	multiport fuel injection
Ignition	electronic coil-on plug	high-energy solid-state direct-fire w/coil near plug
Cylinder block	cast-aluminum w/cast-in iron bore liners	cast-aluminum w/cast-in bore liners
Pistons	cast-aluminum	cast-aluminum
Cam	dual-overhead w/Variable Valve Timing	hydraulic roller-lifter (Active Fuel Management, L99)
Cylinder heads	cast-aluminum	cast-aluminum

Brute force wasn't the 580-horsepower 2012 ZL1's sole claim to fame. "[It] also features exclusive chassis and traction technologies to offer the best of all worlds—including balanced handling for the track, acceleration for a drag strip, and the comfort for a daily driver," said Camaro chief engineer Al Oppenheiser in September 2011. *GM*

(opposite)
A convertible Camaro joined its carryover coupe running mate for 2011. *GM*

Initially offered only as a sport coupe, 2010's Camaro appeared in three trim levels—LS, LT, and SS—with a V-6 standard for the first two, Chevy's 6.2-liter LS3 small-block for the supreme SS. LS3 output in Camaro applications was 426 horsepower, more than enough muscle to put Mustangers in their place, inspiring Ford to trade out 2010's 315-horse 4.6L V-8 for its 5.0L "Coyote" successor, rated at 412 horsepower, in 2011. Bam! Just like that, Detroit's horse race was back on.

Another Camaro birthday was marked in 2012 by the 45th Anniversary Special Edition. Offered for 2LT and 2SS coupes and convertibles, this package featured a Carbon Flash Metallic exterior complemented by a Jet Black interior, 20-inch dark silver wheels, and various bits of commemorative identification inside and out. *GM*

Chevrolet's most powerful convertible ever, a topless ZL1 appeared for 2013 armed, like its coupe counterpart, with 580 horsepower. *GM*

An RS appearance package was available throughout the fifth-gen run, a convertible Camaro debuted for 2011, and thirteen specialty packages came and went during those six years, ending with 2015's Commemorative Edition. "The fifth-generation Camaro was a success from the moment it was introduced," said marketing manager Todd Christensen while announcing this last hurrah in November 11, 2014. "We wanted to create a special edition that celebrates that success as the [latest] generation comes to a close."

Various born-again models also appeared, including the COPO drag car (see "COPO Comeback," page 120) and an awesome 580-horse ZL1, both introduced for 2012. Powered by a supercharged 6.2-liter LSA small-block, the ZL1 ranked as the meanest streetwise Camaro to ever punish a pike. "ZL1 delivers supercar performance and technology in the sports-car segment," said Camaro chief engineer Al Oppenheiser in September

Reintroduced in 2013 for manual-trans SS coupes only, Chevrolet's new 1LE package offered 426 horsepower and 1g cornering for less than $40,000. According to Al Oppenheiser, "the Camaro 1LE is the most affordable, most capable track-day cars offered by any manufacturer." ***GM***

WELL ISN'T THAT SPECIAL...?

Customers' heads were surely spinning as Camaro's fifth generation rolled on, what with so many choices presented to them. Along with the COPO, 1LE, ZL1 and Z28 supercars, there also were thirteen specialty models offered during five of the six fifth-gen years. None were offered in 2014.

2010-15 Special-Edition Camaro Production Figures

Model	2010	2011	2012	2013	2014	2015
Indy 500 Pace Car Replica	294	500	n/a	n/a	n/a	n/a
Transformers Edition	1,916	n/a	1,411	n/a	n/a	n/a
Neimann-Marcus Edition	n/a	75	n/a	n/a	n/a	n/a
Synergy Edition	2,425	103 w/V-6; 255 w/V-8	76, 2LT; 158, 2 SS	n/a	n/a	n/a
Commemorative Edition	n/a	n/a	n/a	n/a	n/a	2,312
45th Anniversary Edition	n/a	n/a	8,283	n/a	n/a	n/a
Camaro Dusk	n/a	n/a	n/a	399, 2LT; 613 2SS	n/a	n/a
Hot Wheels Edition	n/a	n/a	n/a	1,524	n/a	n/a
Green Flash Edition	n/a	n/a	n/a	n/a	n/a	729 coupes, 82 convt.

Chevy's Z/28.R Camaro was introduced at Daytona in January 2014 during the opening race in the IMSA Continental Tire SportsCar Challenge series. "A new chapter in racing begins this week, as the new Z/28 makes its competition debut in one of the most hotly contested series in motorsports," said Chevrolet racing director Mark Kent. "Apart from series-mandated equipment and the specialized suspension components needed for endurance racing, the Z/28.R is as close to a production-spec race car as you'll find." ***GM***

2013–2015 COPO Production Figures

	Engine	Transmission	Paint	CP	Graphics
2013	59 w/427, 7 w/396, 3w/350 (all naturally aspirated)	50 w/auto; 19 w/manual	23 Summit White; 18 Black; 17 Victory Red; 6 Silver Metallic; 5 Ashen Gray	14	62
2014	32 w/350 supercharged; 31 w/427; 4 w/350 naturally aspirated; 2 w/396	61 w/auto; 8 w/manual	25 red; 18 black; 10 white; 8 blue; 8 silver	7	59
2015	29 w/427; 37 w/350 supercharged; 2 w/350 naturally aspirated; 1 w/396	63 w/auto; 6 w/manual	34 white; 14 red; 13 black; 8 silver	9	59

NOTE: *CP is Collector's Package;* *see "COPO Comeback," page 120,* *for 2012 production figures*

Chevrolet officials announced the rebirth of the fabled Z/28 in March 2013. Along with its 505-horsepower Corvette-sourced LS7 V-8, 2014's Z/28 also featured a radically revised chassis, brutal brakes, and a reduced curb weight (300 pounds less than the ZL1) thanks to various lightening touches. *GM*

2011. "For sheer power, it delivers more [horses] than a Ferrari 458, more torque than an Aston Martin DB9 V-12, and a better power-to-weight ratio than a Porsche 911 Carrera GTS."

A ZL1 convertible debuted for 2013 to blow the lid off Chevy's "strongest ever" list. "This is a car that is guaranteed to put a smile on your face every time you drop the top—or hit the gas," continued Oppenheiser. "[It] will be one of the most powerful and most capable convertibles available at any price."

Last seen in 1999, Chevrolet's competition-conscious 1LE package also reappeared in 2013 to make a Camaro SS, again in Oppenheiser's words, "one of the most affordable, most capable track-day cars offered by any manufacturer." "With the 2013 model year, Camaro offers something for almost every driver, including the 323-horsepower, 30-mpg LS; the all-new supercharged ZL1 convertible, the COPO for drag-racing; and the new 1LE for amateur track days," added marketing manager John Fitzpatrick. "We expect

2012 ZL1 Engine	
Type	OHV Gen-IV small-block V-8 (LSA)
Displacement	6.2 liters (376 cubic inches)
Bore	4.06 inches
Stroke	3.62 inches
Horsepower	580 at 6,100 rpm
Torque	556 at 3,800 rpm
Compression	9.1:1
Fuel delivery	sequential fuel injection w/1.9L intercooled Eaton supercharger
Ignition	high-energy electronic
Cylinder block	cast-aluminum w/oil cooler
Crankshaft	forged-steel w/8-bolt flywheel flange
Connecting rods	powder metal
Pistons	hypereutectic cast-aluminum
Cam	hydraulic w/roller lifters
Cylinder heads	cast-aluminum
Valve sizes (inches)	2.16, intake; 1.59, exhaust
Valve lift (inches)	0.480, intake & exhaust
Exhaust system	stainless-steel manifolds

2014 Z/28	
Construction	unitized body/frame; sound deadener and trunk carpeting removed
Model availability	Coupe (w/aerodynamic package & thinned rear window)
Price	$75,000
Wheelbase	112.3 inches
Weight	3,820 pounds
Suspension	double-ball-joint, multi-link struts w/progressive-rate coil springs & direct-acting 0.98-inch stabilizer, front; 4.5-link independent w/progressive-rate coil-over shock absorbers, 1.02-inch stabilizer bar, rear
Steering	variable-rate rack & pinion
Wheels (inches)	19x11, front; 19x11.5, rear
Tires	Pirelli PZeroTrofeo R; P305/30ZR19, front & rear
Brakes	power-assisted 4-wheel discs; Brembo Carbon Ceramic Matrix rotors, front & rear; 6-piston calipers, front; 4-piston calipers, rear
Brake sizes	15.5-inch rotors, front; 15.3-inch, rear
Engine	505-hp 7.0L LS7 V-8
Transmission	6-speed Tremec TR6060 manual
Axle ratio	3.91:1 (w/StabiliTrack electronic traction control)

[this] range of choices, and enhancements, will help Camaro remain America's most-popular sports car."

Choices expanded yet again in 2014 after Chevrolet brought back the Z/28, powered by the naturally aspired LS7 small-block, first seen between Z06 Corvette flanks in 2006. Weighing 300 pounds less than a supercharged ZL1 and able to pull 1.05 g in the turns, the latest, greatest Z truly was, in Chevy's words, "the most track-capable Camaro ever."

"This car restores the mission of the original Z/28, and serves as a testament to the expertise of Chevrolet as the best-selling brand of performance cars," said GM North America president Mark Reuss in March 2013. "The build sheet is the wish list of any racer: lightweight, high-revving, dry-sump LS7 engine; carbon-ceramic brakes; integrated coolers

Featuring RS appearance touches and exclusive green metallic paint, the Green Flash Edition option (RPO B7J) was originally labeled "Spring Edition" when introduced at 2014's SEMA show. Chevrolet then opened an online contest to rename this package, resulting in a more accurate, color-correct tag. The Green Flash package was available only for 2015 SS coupes and convertibles. ***GM***

2014 Z/28 Engine	
Type	OHV Gen-IV small-block V-8 (LS7)
Displacement	7.0 liters (427 cubic inches)
Bore	4.125 inches
Stroke	4.00 inches
Horsepower	505 at 6,300 rpm
Torque	470 at 4,800 rpm
Compression	11:1
Fuel delivery	sequential fuel injection; cold-air induction system w/low-restriction K&N filter
Ignition	high-energy electronic
Cylinder block	cast-aluminum w/pressed-in steel cylinder liners, dry-sump oiling & liquid-to-liquid oil cooler
Crankshaft	forged-steel
Connecting rods	titanium
Pistons	cast-aluminum flat-top
Cam	hydraulic w/roller lifters
Cylinder heads	cast-aluminum, CNC-ported; 70cc combustion-chamber volume
Valve sizes (inches)	2.20, intake; 1.52, exhaust (titanium intakes, sodium-filled exhausts)
Valve lift (inches)	0.594, intake & exhaust
Exhaust system	hydroformed "tri-Y" headers w/dual-mode system featuring 3-inch pipes

(opposite)
Chevrolet offered nine different Camaros for 2015, ranging from the base V-6 LS to the drag-strip-only COPO, far left. ***GM***

for track use; true aerodynamic downforce; and a significant reduction in curb weight. This car could only come from Chevrolet, and could only be called Z/28."

Even with the revived Z-car now running amuck, Camaro found itself again running neck and neck with Ford's Mustang, which was regenerated for 2015 and touting its own accolades. As mentioned, that same year was the last for its fifth-gen rival from Chevrolet. But when the curtain closed this time, it didn't signify the end of the show.

Not by any means.

2010–2015 Production Figures

	LS	LT	SS	ZL1	Z28	Total
2010	7,767	51,469[1]	70,169[2]	n/a	n/a	**129,405[3]**
	[1]included 26,064 w/RS Package [2]included 55,144 w/RS Package [3]all coupes					
2011	14,283	50,634[4]	42,070[5]	n/a	n/a	**106,987[6]**
	[4]included 42,005 coupes, 8,629 convertibles [5]included 29,664 coupes, 12,406 convertibles [6]included 66,272 w/RS Package					
2012	18,085	42,889[7]	27,365[8]	1,971[9]	n/a	**90,310[10]**
	[7]included 33,831 coupes, 9,058 convertibles [8]included 9,058 convertibles [9]included 1,300 w/manual trans, 671 w/automatic (all coupes) [10]included 47,221 w/RS Package					
2013	22,209	40,412[11]	23,855[12]	7,956[13]	n/a	**94,432[14]**
	[11]included 32,344 coupes, 8,068 convertibles [12]included 19,147coupes, 4,708 convertibles [12]also included 1,826 manual trans coupes w/1LE Performance Package [13]included 6,039 coupes, 1,917 convertibles (4,158 w/manual trans, 3,798 w/auto) [14]included 40,510 w/RS Package					
2014	25,674	45,807[15]	23,200[16]	2,436[17]	515	**97,632[18]**
	[15]included 35,961 coupes, 9,846 convertibles [16]included 19,156 coupes, 4,044 convertibles [16]also included 2,468 manual trans coupes w/1LE Performance Package [17]included 1,895coupes, 541 convertibles [18]included 45,392 w/RS Package					
2015	28,934	51,787[19]	21,197[20]	1,850[21]	1,292	**105,060[22]**
	[19]included 37,449 coupes, 14,338 convertibles [20]included 17,190 coupes, 4,007 convertibles [20]also included 2,737 manual trans coupes w/1LE Performance Package [21]included 1,561 coupes, 289 convertibles [22]included 42,625 w/RS Package					

NOTE 1: *All LS & Z28 models were coupes*

Specialty models abounded during the highly successful fifth-gen run, and the era couldn't end without one final fling. "We wanted to create a special-edition Camaro that celebrated that success as the fifth generation comes to a close," said Camaro marketing manager Todd Christensen in November 2014. The 2015 Commemorative Edition included the RS package option and was offered for 2LT and 2SS coupes and convertibles in five exterior color combos. Exclusive striping, 20-inch wheels, a ZL1 rear spoiler, and special identification also were part of the deal. *GM*

BEST YET

10

Chevy's Latest/Greatest Camaro

"**When the fifth-generation Camaro arrived** as a 2010 model, it finally banished the [breed's] image as a crude muscle car," wrote veteran journalist Csaba Csere in *Car and Driver's* March 2015 edition. "With its independent rear suspension, solid structure, and strong powertrains, the [2010] Camaro was thoroughly modern. But it [also] successfully incorporated cues from the [Sixties] original and was instantly recognizable, and appreciated, by Chevy enthusiasts."

That cover-all-bases aspect, appealing to both old friends and newfound fans, helped Chevrolet's latest Camaro do what no preceding pony had done since 1991: leave its rival from Dearborn in the dust. By 2015, Camaro stood as Detroit's best-selling performance car for five year running (by calendar count; Mustang was tops in model-year numbers for 2012 and 2014), and total fifth-gen production surpassed the half-million mark in March that year. Good work, no?

Sure, but recall the comment inked on an earlier page, the one concerning proverbial laurels and how unwise it is to plant your rump on 'em. Ford folks, mind you, weren't sitting on their brains while Camaro was rearranging the

(opposite)
Ford people barely had time to celebrate their new sixth-generation pony car in 2015 before the Bowtie gang came right back with their all-new sixth-gen Camaro, *Motor Trend*'s Car of the Year for 2016. The Chevy in front is a 2016 LT; the droptop Mustang in back is a 2015 50th Anniversary GT.

sales race, and the radically redesigned 2015 Mustang sure looked like it might revive past runaway memories, especially if Chevrolet didn't whip up something new, pronto.

Consider it done.

Work on a new-and-improved sixth-generation Camaro began at November's end in 2011, with Chevrolet Performance Car studio staff first determining basic architectural proportions. A DSI (Document of Strategic Intent) meeting followed on March 5, 2012, effectively kicking of the design process, which, by Ed Welburn's invitation, was open to every GM studio on this planet. "Who wouldn't want to have a hand in [creating] one of the industry's most iconic vehicles," said Chevrolet Performance Car Exterior Design director Tom Peters two years later. "We had an outstanding response with a wide range of ideas from around the world."

Welburn's worldwide contest was among the keys to making the latest, greatest model all it could be. "This process was quite unique," said Peters in March 2016. "Ed always sets up competition between groups. I equate it to competitive swimming. If you're going for fast times by yourself, sure, you can do well. But if you have your buddy pushing you in the next lane, you do even better. Ed is well aware of the value of this kind of competition."

By the end of April, a couple hundred submissions were whittled down to one exciting sketch, done back in Warren, Michigan, by Chevrolet Performance Car's Hwasup Lee, who also was responsible for 2014's Corvette Stingray. What was it like to be the main man behind the machine in both blood-pumping cases? "I feel very proud," he said in March 2016. "It was a really special experience, a dream come true. But I got a lot of help, and there was a lot of pressure, too. I thought, 'I can't screw up these cars.'" Trust us; you can open your eyes now, Hwasup.

As usual, taking pre-production sixth-gen Camaros out in public for real-world testing required artful camouflage to hopefully keep this cat—err, pony—in the bag. GM

As far as Camaro was concerned, Lee's guiding light once again shined back to the 1960s before focusing on the future. "The mission was to further refine the iconic Camaro, maintain the heritage of the vehicle and create a new and fresh statement while ensuring that anyone who sees it knows it's a Camaro," he said. "Although every phase of this program was challenging, creating a new spirit of Camaro was a rare opportunity for a designer to have a blank canvas to [draw up] an all-new vehicle."

Lee's idea was first translated into a ⅓-scale model, which then morphed into various full-size clays for GM's traditional "patio review," held outside its Warren studio on July 9, 2012. The first STD (Single Theme Direction) gathering, with Welburn presiding, occurred on November 8, leaving two final competitors (out of seven considered) for a continuing "theme bake-off." GM's Global Design vice president wound down the Single Theme process on December 19, then made his ultimate choice official during the DSO (Design Sign Off) event, held May 16, 2013. His bosses atop the executive leadership ladder added their approval in June.

Global Design head Ed Welburn invited GM studios from around the world to submit sketches for the sixth-gen Camaro project. South Korean–born Hwasup Lee's drawing was chosen from hundreds of submissions. *GM*

Prototype testing began in April 2014, and Chevrolet's official announcement of the sixth-gen Camaro's public unveiling followed on March 16, 2015. About a thousand people attended that gala, held at Detroit's Belle Isle Park, on Saturday, May 16, 2015. "We wanted to share the moment with the customers and fans who have helped make Camaro both America's favorite performance car for

Refining the 2016 Camaro's final shape in clay typically required more than six months of dedicated artistry. *GM*

The 2016 Camaro design team (left to right): Adam Barry, Casey Swanseger, Brett Golliff, John Mack, Tristan Murphy, Ryan Vaughan, Hwasup Lee, Jason Ross, Kirk Bennion, Tom Peters, Brian Malczewski, and Rich Scheer. Team members not shown here: Chevrolet Global design executive Ken Parkinson, Helen Emsley, Scott Wassel, Mark Ferri, Roch Pnakovich, Kelly Craigo, James D Ciolfi, and Brianne Hoffman. ***GM***

the past five years and a cultural icon since 1967," said marketing manager Todd Christensen.

"We've taken that iconic design and amplified its proportions to reflect a more dynamic driving experience—like the T-shirt on a muscular physique," added Tom Peters in June 2015. In *Car and Driver*'s October 2015 issue, Alexander Stoklosa described the 2016 Camaro SS as a "binge eater of roads." He went on to say that "this is a Camaro that finds itself closer to the sports-car end of the spectrum than ever before." "Overjoyed" might be a fair portrayal of Stoklosa's true feelings at the time concerning the car's mechanical aspects.

Holding everything together this time around was GM's weight-saving, yet still brawny, Alpha platform, originally introduced for Cadillac's compact ATS unveiled in August 2012. However, no one should be fooled: the 2016 Camaro was in no way just a re-bodied Caddy.

"Alpha provided a strong foundation," explained GM Global Product Development vice president Mark Reuss in March 2015. "But more than 70 percent of the components are unique to the Gen 6 Camaro, including

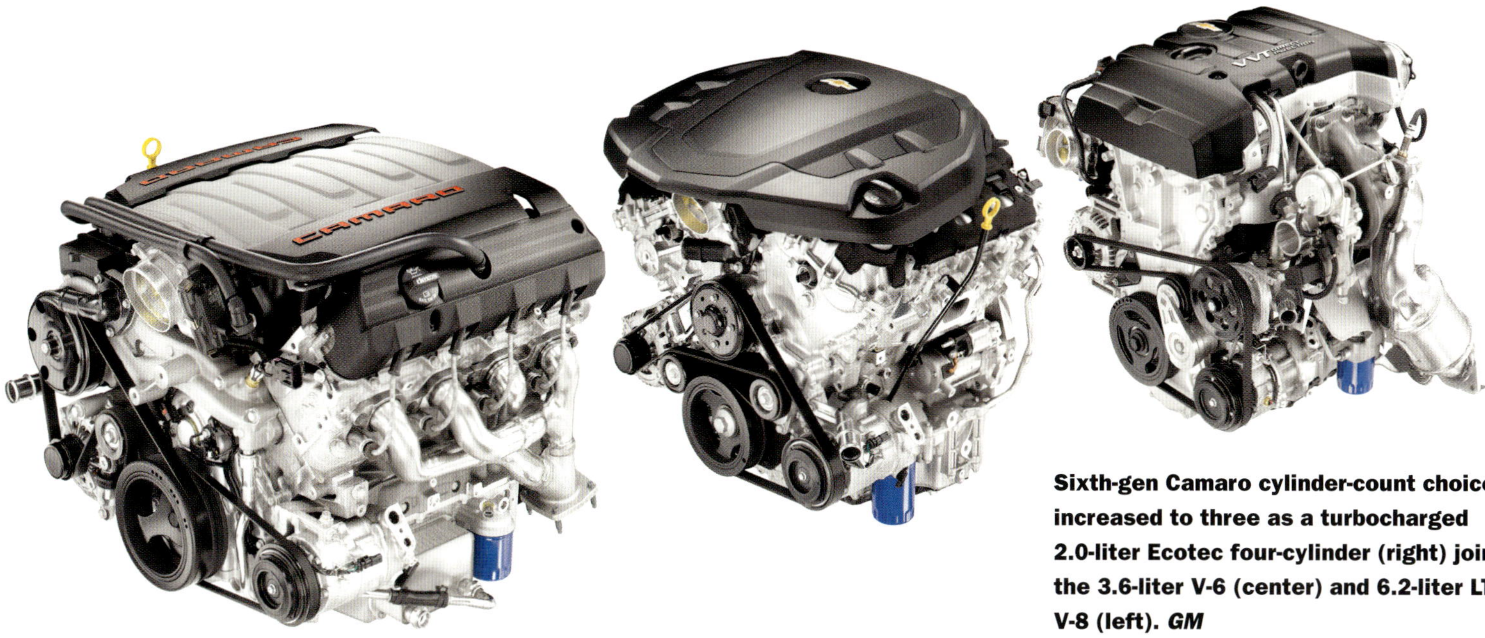

Sixth-gen Camaro cylinder-count choices increased to three as a turbocharged 2.0-liter Ecotec four-cylinder (right) joined the 3.6-liter V-6 (center) and 6.2-liter LT1 V-8 (left). *GM*

exterior and interior dimensions, an all-new interior, front and rear suspension, and powertrain components."

Added Stoklosa, "The Zeta platform, for example, shared by the full-size Pontiac G8 and Chevrolet Caprice sedans, gave the fifth-gen car lane-filling girth and unnecessary heft. The 2016 Camaro SS once again enjoys a close relationship with GM's best products, only this time it networked better. [Its] weight loss can't go overlooked; Chevrolet says the switch to the Alpha bones saves 223 pounds over last year's SS manual coupe for a lithe 3,685-pound curb weight."

Chevrolet initially offered the sixth-gen Camaro in coupe form in three flavors, with the first featuring a new four-cylinder, a 275-horsepower turbocharged Ecotec mill that displaced 2.0 liters and was capable of more than 30 miles per gallon while also producing 0–60 clockings in less than six seconds. No, another Iron Duke it wasn't. Next up the list was a 3.6-liter V-6 that made 335 horses—from (for the metric-conversion-challenged) only 222 cubes. Again, the model reference for these Camaros of lesser cylinder count was "LT," and trim levels included 1LT and 2LT, along with 1SS and 2SS.

2016 Camaro	
Construction	unitized body/frame
Model availability	coupe, convertible
Base price	$25,700, 1LT coupe; $29,800, 2LT coupe; $32,700 1LT convertible; $36,800, 2 LT convertible; $36,300, 1SS coupe, $41,300, 2SS coupe; $43,300, 1SS convertible, $48,300, 2SS convertible
Wheelbase	110.7 inches
Curb weight	3,760 pounds (SS coupe)
Suspension	MacPherson-type strut w/dual lower ball joints, twin-tube struts & direct-acting stabilizer, front; independent five-link w/twin-tube shocks & direct-acting stabilizer, rear
Steering	ZF rack & pinion w/variable-ratio electric power assist
Wheels	18-inch, LT; 20-inch, SS
Tires (LT)	Goodyear Eagle Sport all-season, P245/50R-18
Tires (SS)	Goodyear Eagle F1 Asymmetric 3 run-flat; P245/40ZR-20, front; P275/35ZR-20, rear
Brakes	4-wheel discs w/4-channel ABS
Brake sizes (LT)	12.6-inch rotors, front; 12.4-inch, rear; 4-piston calipers, front; single-piston sliding calipers, rear
Brake sizes (SS)	13.6-inch rotors, front; 13.3-inch, rear; 4-piston fixed calipers, front & rear
Engine	275-hp 2.0L turbocharged 4-cylinder; 335-hp DOHC 3.6L V-6; 455-hp OHV 6.2L V-8
Transmission	6-speed Tremec TR3160 manual, std w/LT; 6-speed Tremec TR6080 manual, std w/SS; 8-speed Hydra-Matic BL45 automatic, available w/LT; 8-speed Hydra-Matic BL90 automatic, available w/SS
Axle ratio	3.27:1, std w/manual-trans LT & 2.0L automatic; 3.73:1, std w/ manual-trans SS; 2.77:1 included w/automatic 3.6L V-6 & automatic 6.2L V-8

A trendy flat-bottomed steering wheel was just one of many ergonomic improvements made to the driver-focused sixth-gen interior. ***GM***

At the top, the aforementioned SS was fitted with Chevy's unstoppable Gen-V V-8, introduced two years prior for the first C7 Corvette. Pumping out just as many horses as its Stingray counterpart, 455, Camaro's 6.2-liter LT1 not only offered customers more power and better fuel efficiency at the same time, it also put the "high" in high tech like no other pony car powerplant before.

"The Gen-V program thoroughly modernized the entire engine family," said Small Block global chief engineer and program manager Jordan Lee in January 2016. "Without question, with direct injection, cylinder deactivation, and a very sophisticated control system, our Gen-V [V-8s] are the most advanced engines on the market."

2016 Camaro Engines

Type	DOHC 4-cylinder	DOHC V-6	LT1 OHV V-8
Displacement	2.0 liter (122 cid)	3.6 liter (222 cid)	6.2 liter (376 cid)
Bore	3.39 inches	3.74 inches	4.06 inches
Stroke	3.39 inches	3.37 inches	3.62 inches
Horsepower*	275 at 5,600 rpm	335 at 6,800 rpm	455 at 6,000 rpm
Torque*	295 at 3,000-4,500 rpm	284 at 5,300 rpm	455 at 4,400 rpm
Compression	9.5:1	11.5:1	11.5:1
Fuel delivery	direct high-pressure fuel injection w/turbocharger	direct high-pressure fuel injection	direct high-pressure fuel injection
Cylinder block	cast-aluminum w/cast-in iron bore liners	cast-aluminum w/cast-in iron bore liners	cast-aluminum w/cast-in iron bore liners
Pistons	cast-aluminum	cast-aluminum	cast-aluminum
Cam	dual-overhead	dual-overhead	hydraulic w/roller lifters
Cylinder heads	cast-aluminum w/4 valves per cylinder & continuously variable valve timing	cast-aluminum w/4 valves per cylinder & continuously variable valve timing	cast-aluminum w/2 valves per cylinder & Active Fuel Management

**SAE Certified*

Transmission choices behind all three engines numbered two: a six-speed Tremec manual and Chevy's new eight-speed Hydra-Matic paddle-shift automatic. An 8L45 Hydra-Matic, designed with more efficiency in mind, was installed behind the turbo four and V-6, while an aggressive 8L90 eight-speed was bolted up with a vengeance to the LT1. Automatic-trans V-6/V-8 models also were treated to Active Fuel Management, which disabled cylinders (two in the former's case, four in the latter's) during light-load operation to enhance economy in those applications.

LEADING MAN: HWASUP LEE

South Korean–born Hwasup Lee was a youngster when he determined automotive design work was his future, this after discovering that grownups could actually get paid for drawing cars. He moved to the United States. while in junior high school then completed a transportation design degree at Southern California's Art Center College of Design in 2000, joining General Motors immediately thereafter. GM's truck studio was his first home, followed by positions in Warren, Michigan, and in England. By 2007, he was lead exterior designer for a global crossover platform marketed on this side of the planet as Buick's Encore and Chevrolet's Trax. Three years later, he joined Tom Peters's Performance Car studio, where he led the exterior design crew responsible for the ZL1 Camaro, introduced for 2012. Next came his "winning" sketch for 2014's all-new seventh-generation Corvette, a stunning form that lead the way for his second big score, 2016's sixth-gen Camaro. Suffice it to say that Lee feels his contributions to *both* the C7 Stingray and Chevy's latest pony car represent truly tasty icing on his career cake. "It has been a really special experience, a dream come true." he said in March 2016. And how does he feel whenever he finds himself in traffic with one of these head-turners to either side of him? "By the end of three years, after looking at these drawings so many times, going over it again and again, it kind of feels like it's time to move on. But when you see the actual cars being built, it sure seems like, wow, this is really happening. That's when it all becomes 'worth it.'" As humble as design geniuses come, Lee also still sounds a bit dumbfounded about how lucky he is to have transformed a childhood dream into a rewarding profession. "I really feel honored that I get to work for the legends who are still around here. Tom Peters and John Cafaro are like walking legends." May Hwasup Lee live a thousand years—and never age another day.

Still more than humble after fifteen years on the job at GM, Hwasup Lee came out first twice during Ed Welburn's recent sketch competitions, drawing up early forms for both the C7 Corvette and sixth-generation Camaro. *GM*

SOCIAL ORDER

The Society of Automobile Engineers was founded in 1904 to allow this country's early automakers to share technical knowledge, to foster growth, and progress as a whole. Electric car pioneer Andrew Riker was the SAE's first president, Henry Ford its first vice president, and among the group's initial endeavors was the regular publication of valued technical papers, soon widely known as SAE Journals. Membership was 1,800 by 1916, the year the shingle morphed into Society of *Automotive* Engineers in order to encompass all powered vehicles, be they fitted with wheels, wings, or otherwise. SAE International then became the official label in 2006. Overseeing advertised horsepower ratings has long been among the group's main focuses, but official outputs remained more hit than miss until 1971, when a new SAE standard (coded J1349) specified the use of net figures, not the questionable gross numbers seen up to that point. Though more accurate than before, Detroit's net ratings still left much to be desired as far as honesty was concerned, yet they remained the status quo for another three decades. In April 2005, SAE officials announced new test standards intended to improve the accuracy of advertised output labels. Closing up J1349 loopholes was job one, meaning engine tests now produced real real-world results, not tallies only possible under perfect lab conditions. Adhering to revised J1349 testing, however, was left up to the individual companies; such practices weren't required. Nor was compliance with a totally new standard, J2723, which brought in an independent third party to witness said trials, culminating in SAE certification for an engine being put to the question. Chevrolet's LS7 small-block was Detroit's first motor to undergo the J2723 process. When this Corvette V-8's label read "SAE Certified," customers could be confidant they were taking delivery of all 505 of those promised ponies, an assurance still lacking from some rivals. "It's an extra step not all OEMs (original equipment manufacturers) do, even today,"explained GM Global Propulsion Systems Communications man Tom Read in January 2016. Now, when someone says "Genuine Chevrolet," they mean it more than ever.

Offering the same SAE-certified horsepower count (455) as its Corvette counterpart, Chevrolet's LT1 Gen-V small-block made 2016's SS the most powerful Camaro ever. ***GM***

Yet another choice was announced in June 2015, this one thrilling both LT and SS customers who preferred driving topless. According to Chevrolet, the 2016 Camaro convertible was the only open-air model in the segment to feature all of the following in combination: full automatic operation (including latches), remote (key fob) opening, open/close capability up to speeds of thirty miles per hour, and a hard tonneau that also went into place without any help from human hands.

"With many convertibles, you have to affix a tonneau cover manually—if it's done at all," said Tom Peters while helping introduce the sixth-generation ragtop. "The Camaro convertible's automatically deploying tonneau not only makes it easier to

2017 1LE Package

Camaro LT coupe	Camaro SS coupe
3.6L V-6 w/6-speed manual-trans	6.2L V-8 w/6-speed manual-trans
FE3 suspension (from SS); included dampers, rear cradle mounts, ball-jointed rear toe links & stabilizer bars	Magnetic Ride Control
lightweight 20-inch forged-aluminum wheels w/Goodyear Eagle F1 tires (P245/4040R-20, front; P275/35R-20, rear)	segment-exclusive electronic limited-slip differential w/3.73:1 ratio
Brembo 4-piston front brake calipers	unique forged-aluminum wheels &Goodyear Eagle F1 Supercar tires w/compound & construction developed exclusively for Camaro (P285/30ZR-20, front; P305/30ZR-20, rear)
mechanical limited-slip differential w/3.27:1 ratio	Brembo brakes w/6-piston monobloc red calipers & 2-piece 14.6-inch rotors, front
track-cooling package w/oil, differential & transmission coolers	track-cooling system w/oil, differential & transmission coolers
suede steering wheel & shorter-throw shifter	suede steering wheel & shorter-throw shifter
dual-mode exhaust system	dual-mode exhaust system
SS fuel system for higher-load cornering capability	Recaro front seats w/aggressive bolsters optimized for shifting & steering comfort
satin-black hood, front splitter & unique	satin-black hood, mirrors & unique
3-piece rear spoiler	3-piece rear spoiler
unique high-flow upper & lower grille w/satin-black accents	Satin Graphite front splitter & rear diffuser
available Recaro seats & PDR video/data recording system	available PDR video/data recording system

"From the beginning, Camaro's architecture was developed to incorporate a convertible with uncompromised driving dynamics," said chief engineer Al Oppenheiser while announcing the coupe's topless running mate in June 2015. "Customers will appreciate what they don't feel: quivers, cowl shakes or an under-damped chassis typically found in a four-seat convertible." ***GM***

C. M. MILLICAN
BLACKSMITHING
EST. 1909

enjoy convertible driving when the inspiration hits, it ensures the car always looks its best."

Of further note was the ride encountered once that top disappeared. "From the beginning, the [sixth-gen] architecture was developed to incorporate a convertible with uncompromised driving dynamics," added Camaro chief engineer Al Oppenheiser. "Customers will appreciate what they don't feel: quivers, cowl shakes or an under-damped chassis typically found in a four-seat convertible."

The chassis also received an upgrade in 2017 after Chevrolet reintroduced its latest 1LE option, this time for both V-8 SS and V-6 LT Camaros. In each case, the installation will be limited to the manual-trans coupe, a perfect choice for rockin' and rollin' around a road course.

"The 1LE packages follow a recipe any track-day enthusiast will appreciate," said Oppenheiser while announcing the latest variation on the sixth-gen theme in February 2016. "Start with a lightweight, stiff structure and then add higher levels of suspension stiffness and damping, more aggressive tires, higher downforce, more powerful brakes, and more cooling. The result delivers better performance in all measures. In our tests, the V-6 Camaro 1LE delivers lap times comparable with the previous-gen V-8 model."

(opposite)
The 2016 pecking order consisted of the LT (available with the turbo four-cylinder or V-6) and the V-8 SS. Two trim levels (1LT, 2LT; 1SS, 2SS) again were offered in each case. Shown here is a V-6 LT adorned with the optional RS appearance package.

Chevrolet brought the 1LE performance package back for 2017's Camaro, but this time the deal also was available for V-6 LT coupes (demonstrated at left) along with V-8 models, exemplified at right. *GM*

2017 50th Anniversary Edition
Nightfall Gray Metallic exterior (black top w/convertible)
Specific 20-inch 50th Anniversary wheels & center caps
RS Appearance Package included w/2LT version
Specific 50th Anniversary stripe package & exterior badges
Unique grille w/satin chrome accents
Body-colored front splitter
Orange brake calipers (front only on LT)
Unique black leather interior w/suede inserts & orange accent stitching
Distinct 50th Anniversary treatments on instrument panel, seatbacks, steering wheel & illuminated sill plates

Now thoroughly impressed, Alexander Stoklosa couldn't say enough good things about what the new 1LE option did to transform the already buff sixth-gen Camaro, especially the six-packer, into a seriously ripped hunk. "With the package [now] available on the 1LT V-6, it might be possible to snag a 1LE for about $30,000," he wrote in *Car and Driver's* February 2016 edition. "If that's not reason enough to drop the 'V-8 or die!' line, we suggest you go test-drive a six-cylinder Camaro right now. We'll try not to say we told you so."

A real no-brainer, a 50th Anniversary Edition package, was announced in March 2016 as of this writing. Available for 2LT and 2SS Camaros with or without tops, this commemorative treatment among other things included Nightfall Grey Metallic paint, specific 20-inch wheels, and all the appearance stuff you'd expect to help mark a half-century-old heritage. Not long afterward, GM also announced that a 50th Anniversary coupe would pace the 100th running of the Indy 500 in May 2016. Again, as expected.

A half-century's worth of Camaro history was marked in March 2016, when Chevrolet announced its own 50th Anniversary pony car. *GM*

How many new models, not to mention model years, await the faithful once those fifty candles burn out is anyone's guess. "Team Camaro never sleeps, we're always working on something," said the ever-doting Fbodfather, Scott Settlemire, when pressed in February 2016 to peek into a crystal ball. "If you think the SS and 1LE are hot, you ain't seen nothing yet. There's always something ahead."

"I love my '69 Camaro, and I treasure it," added Ed Welburn, also in February 2016. "[But] I would really rather drive the [sixth-gen] Camaro with its stiff body structure, incredible suspension, refined interior, current safety features, and an exciting design." How soon will he say something similar about the present car? "I fully expect Camaros of the future to be just as surprising, with equally exhilarating designs and advance technologies that are hard to imagine today."

Hard to imagine? Go ahead, Ed, give us a try.

Motorsports legend Roger Penske drove the 2017 Camaro SS 50th Anniversary Edition pace car at the one-hundredth running of the Indianapolis 500. *GM*

INDEX